BTEC Nationals – IT Practitioners Core Units Computing and IT Tutor Resource Pack

T0231346

BTEC Nationals – IT Practitioners Core Units for Computing and IT Tutor Resource Pack

Sharon Yull

Howard Anderson

Routledge
Taylor & Francis Group

LONDON AND NEW YORK

First published 2003 by Newnes

First edition 2003

Published 2021 by Routledge
2 Park Square, Milton Park, Abingdon, Oxon OX14 4RN
605 Third Avenue, New York, NY 10017

Routledge is an imprint of the Taylor & Francis Group, an informa business

ISBN 13: 978-0-7506-5687-0 (pbk)

Composition by Scribe Design, Gillingham, Kent

British Library Cataloguing in Publication Data
A catalogue record for this book is available from the British Library

Contents

 # Introduction

This tutor's resource pack has been designed to provide activities for students to *do* rather than simply supply material extra to the book.

One of the main difficulties in preparing the book and this pack is in the setting of the correct level. The BTEC National requirements are a good guide but often examples are required to add clarity. Computing is now such a large subject that no treatment at level 3 can possibly hope to cover every aspect; some aspects will necessarily be given more prominence than others. This pack will concentrate more on the items that require practice such as programming and communications, and less on sets of facts to be remembered.

It has been recognized that some aspects of the BTEC Unit requirements are catered for using centre-specific equipment and software. For example, some centres may use Apple computers, others may use PCs with Linux, others still will use PCs running any of the several versions of the Microsoft Windows operating system. For this reason, areas that are likely to be centre specific will not be found in this pack. In particular, practical exercises to cover the operating systems and networking parts of the units will have to be centre specific. Wherever possible, the tasks in this pack are generic in nature except the Pascal answers. These were prepared using Free Pascal that runs in PCs.

The pack is designed to provide the following support tools:

- Exercises and student tasks
- Case study material
- Group activities
- Research activities

All worksheets have been prepared in conjunction with the BTEC IT Practitioners book and complement the material provided under the given chapter headings.

The worksheets have been designed to enable both the tutor and the students access to material that is current, relevant and appropriate to this level of learner. The majority of worksheets can be easily photocopied or transferred onto transparencies.

The worksheets have been designed to test prior knowledge, understanding and also to consolidate learning of a certain topic area.

The answers to each worksheet are examples only. Students should be encouraged to seek their own style so the answers in this pack should be used as a guide only. Programs and other testable items have only been tested to a standard appropriate to the course, i.e. not with the same rigour that would be applied to commercial software.

For 'fact finding' aspects of the units, students should be encouraged to use the internet to the full. As computing moves at a great rate, only the internet can give the latest information. One year the fastest PC available had a processor fitted that ran at 800 MHz, just a few months later, speeds of 1.2 GHz were available and at a lower cost than the 800 MHz machines. As this text is being prepared, machines of 1.7 GHz are common place and 2.4 GHz are available, but by the time the pack is printed and in use, far faster machines will be available.

For this reason, one of the most useful skills that students should develop is use of the internet. It is not often realized that successful use of search engines etc. depends more on the user's knowledge of the subject and their ability in the use of language than plain 'computer skills'. A historian is more likely to find good historical information than a 'computer whizz'.

Sharon Yull
Howard Anderson

Unit 1 Language and communications

1.1 How do we communicate?

Worksheet 1

Tools for communication

To enable effective communication there needs to be three main elements: a sender, a communication tool and a receiver.

1. Using a similar template to the one provided, give an example of a sender, communication tool and receiver for the following systems:
    (i)   Receiving information about a course at college
    (ii) Booking a holiday
    (iii) Invitation to a party
    (iv) Obtaining information about computer hardware for an assignment
2. Identify five factors that could impact upon the way in which information is transmitted between a sender and a receiver.
3. Identify five communication tools and provide a comparison of each in terms of:
    (i)   Speed of transmission
    (ii) Cost of transmission
    (iii) Reliability of transmission
    (iv) Formality of transmission

Figure 1.1 *Elements to enable effective communication*

Worksheet 2

Impact of communication tools

Different types of communication tools are used on a daily basis to transmit information. The type of tool that we use is dependent upon a number of factors, including:

- Time
- Expense
- Geography
- Impact
- Convenience
- Audience
- Occasion
- Formality

1. Complete the table identifying two suitable communication tools for each given scenario justifying your choice fully.

Scenario	Choice 1	Choice 2
You need to send information which is personal and confidential to a friend who lives abroad		
You promised that an assignment would be handed in today, the last possible hand-in date		
You have just passed your driving test and cannot wait to tell your friends and family		
You have seen a suitable part-time job advertised in the paper and require further information		
You have been stopped in the street by an exchange student who speaks little English and needs directions to the bank		

1.2 Types of communication

Worksheet 3

Verbal communication

It has been established that for effective communication to take place there need to be three components: the sender, use of a communication tool and a receiver. However, another factor that will influence the effectiveness of communication is the type of communication that is used.

Verbal communication is one of four categories of communication that are used to transmit information. The way in which we transmit information verbally depends upon who we are speaking to, the environment and the objective of the transmission.

1. Complete the following table by identifying a situation that you have been in which involved that particular category of verbal communication.

Categories	Situation
Negotiating	
Persuading	
Debating	
Delegating	
Challenging	
Advising	
Arguing	
Apologizing	

Worksheet 4

Impact of verbal communication

The impact of verbal communication can be very strong, and as a result change your relationship with the person/people you are speaking to momentarily or permanently.

1. Think of a situation where you have been the receiver of verbal communication recently and think about how you felt towards the receiver under the following situations:
    (i)   Being told off or disciplined
    (ii) Being congratulated
    (iii) Being advised
    (iv) Being confided in

Verbal communication is probably the most important and most used form of communication. As a result verbal communication has a great deal of benefits over other communication methods.

2. Identify six advantages of using verbal communication.
3. Identify any limitations of using verbal communication.

Interviews are a good example of testing how well you communicate verbally because you have to present yourself professionally and competently. Your skills of trying to persuade and convince the interviewer that you are the best candidate for the job are required from the moment you step in to the moment you walk out.

You are attending an interview for a part-time job at your local library. The position requires somebody to work as part of a team to help update the computer systems and make sure that all new catalogue titles are stored electronically. The job description also mentions somebody with software skills, especially databases and web-based experience.

4. Draw up a list of topics that you feel you could discuss at the interview in relation to your own experiences and skills.
5. What further information might you require at the interview?

1.3   Written communication

Worksheet 5

Formal versus informal written communication

Written communication can be divided into two areas, formal and informal. Formal written communication relates to official documents which provide guarantees and assurances, these documents can be legally binding. Informal written communication includes letters to friends, memos, e-mails and greetings etc.

1. For each of the given scenarios identify what type of written document could be used and state whether or not it is formal or informal.
   (i) A managing director provides information about the performance of the company over the last quarter at a board meeting
   (ii) A colleague from the sales department needs to send information about the team briefing to other department members
   (iii) The personnel department informs applicants as to whether or not they have been successful at the interview
   (iv) Confirmation is provided that the transaction of your buying concert tickets is complete
   (v) Information is required about the current status of your bank balance
   (vi) Information is provided on your first day at work about your responsibilities, terms and conditions of employment
2. Identify three other formal and three other informal documents (do not specify any given for task 1).
3. Why is it important to have both a formal and informal document?

Worksheet 6

Producing written documents

The following are samples of documents that are used in organizations:

- Letter
- Memorandum
- Report

1. Using similar templates produce each of the documents using applications software (do not use the wizard function).
2. Once the templates have been set up and saved with the appropriate headings for inserting dates, addresses, subject headings and closure sections print out a copy of each.
3. Using the information given complete the template documents and save the new documents as different files.

Letter content

A formal letter needs to be sent out to:

Mr and Mrs Richard Wright
36 Honeypot Lane
Colney
Norfolk
NR12 4ER

The letter is from:

Norfolk Gardening Society
The Priory
Cheddar Way
Norfolk
NR4 9LO

Date: Friday 3 May 2002

A suitable introductory heading is required to open the letter.

Mr and Mrs Wright have won first prize in the Norfolk Gardening competition. They have won £250 of gardening vouchers and a plaque which will be presented to them in June at the Society's annual dinner dance.

Invitations are enclosed, reply requested ASAP.

Mel Gladding
Chair of the Society

Memorandum content

A memo needs to be sent to Michael Peterson, director of finance, and copied to Paul Graham, Tim Phillips, Amanda Greene and Jenny Crop (all finance administrators). Dated 22 March 2002. The memo is regarding the team building weekend.

Just to remind you all that the finance team building weekend is taking place this May Day Bank Holiday. Can we all arrange to meet outside the foyer at 10:00 where transportation will be waiting. Walking boots are essential.

Report content

You have been asked to prepare a report as part of your Language and Communications unit on the following topic area.
    'Computer gaming has changed radically over the years from an isolated activity to a social and interactive pastime.' Discuss.
    For the report you will need to research the following areas:

(i) Over a set period of time, e.g. 10 years, identify how computer gaming has changed in terms of hardware and software developments and technology.
(ii) Describe recent developments in the areas of: games consoles, online gaming, interactive software and interactive facilities such as companies offering these facilities.
(iii) Gather statistical figures about sales of games consoles.
(iv) Conclusions can be based on personal opinions about the social and interactive aspects of computer gaming. Social implications of computer gaming, e.g. too much time wasted, or the future of computer gaming.

Sample letter

Sender's information:
Mr Spencer James
4 Toad Cottage
Armley
NR32 4DD

If there is no letterhead the address could go to the right-hand margin: ⟶

Recipient Information

The Royal Aircraft Club
Highbury House
Staunton
Essex
CO31 7JD

Date:
29 April 2002

Reference number: (if applicable)

Salutation: For the attention of the Company Secretary

Introduction: Renewal of club membership

Content:

_____

_____

_____

_____

_____

Closure:

Yours faithfully

Spencer James

Sample memo

Memorandum

To: Jean – Sales Manager

From: Carol – Marketing Manager

CC: Mark – Marketing Director

Date: 22/02/2002

Re: Launch of new marketing campaign

Body of text would be displayed here:

Just a reminder that the room for the marketing campaign meeting has changed from 1b to 2d on the second floor.

Don't forget the biscuits!

Sample report

Marketing campaign
2002

Strategies for implementation

Jean Dye

22 February 2002
Commissioned for Sales and Marketing

Contents

1.0 Introduction: the introduction should provide a short summary of the overall focus and content of the report.

2.0 Procedures: identification of any procedures used to collect, collate, analyse and present information.

3.0 Main findings: the main findings section is where the bulk of the report content should be placed. The main findings section should be broken down into task, action or research areas. Each area of the findings section should put forward arguments or statements supported by research and analysis. The main findings section can be broken down further into subsections, for example:

3.1 Marketing campaign
    3.1.1 Promotional activities
    3.1.2 Dates of launch
    3.1.3 Target audiences

4.0 Conclusions: the conclusion section should bring together all of the items discussed within the main findings section and provide a summary of the key areas identified.

5.0 Recommendations: this section is solution based, providing the subjects of the report with proposals as to how they can move forward with the report objective.
For example, recommendations for promotional activities could include:
(i)   Getting prices for local radio air time slots
(ii)   Identifying competitor strategies for marketing of the particular product
(iii)   Set up launch days for the new product

6.0 References: this section should identify and give credit for all information sources used to include books, magazines or journals, other documents or reports, and the internet etc.

Appendices: this section will provide supporting documentation to give additionality to the report content. Appendices could include lists of facts and figures, leaflets, downloaded information, photocopied material etc.

Worksheet 7

Agendas

Agendas are used to provide a structure of activities or items to be discussed at a meeting. An agenda is constructed from the intended contributions of people attending the meeting. A set format is used similar to the example provided giving a list of items to be addressed.

AGENDA
Marketing Campaign

22 February 2002
Room 2d second floor

1. Background on product to be launched
2. Predicted sales figures
3. Proposals for marketing campaign
   • Radio air time
   • Billboard
   • Magazine/newspaper advertisements
   • Launch days
4. Resources
5. Costs
6. Time scale
7. Strategies for implementation
8. Any other issues

1. Prepare two agendas based on the following topic areas:
   (i) Planning a group holiday with friends
   (ii) Setting up a weekly film night at your school/college
2. Each agenda should have at least six points for discussion.
3. For three points listed state what will be discussed more specifically by breaking it down (as shown in item 3 of the agenda above).

Worksheet 8

Creating a CV

A curriculum vitae is an important document as it provides you with a detailed summary of your qualifications, skills and achievements (work based, academically or socially). CVs are used to complement applications for a job or entrance into higher education or training.

There are a number of guidelines about what a CV should have on it, its layout and length, but at the end of the day a CV should provide facts and information about your abilities. Generally a CV should be no longer than two A4 sheets; some stipulate only one, but this can be difficult if you are experienced and qualified in a number of areas.

A CV is personal and it should be unique to you; some people have different CVs promoting different skill aspects depending upon the application. One thing to remember is that your CV may end up in a pile with another hundred – what will make them read yours?

1. Based on the template headings provided produce an up-to-date CV. Try to include information under each heading (if you have never had a job, what about including work experience from school?).
2. Write down a list of all of your skills and abilities, identify what you have achieved. You could produce one academic CV and one work-based CV.

Curriculum vitae

Name
Address
Contact details (telephone and e-mail address)
Date of birth

Education and qualifications

| Secondary school details | Qualifications | Dates | |
| To: | | From: | |

Further education details   Qualifications   Dates
                                                           To: From:

Other certificates and/or qualifications:

Work experience

Employer details   Responsibilities   Dates
                                                     To: From:

Skills and abilities

For example applications software, a list of packages and programming languages that you can use, state whether or not you have developed or designed anything, e.g. database, website etc.
Passed driving test, team captain, voluntary work etc.

Hobbies

Try to put down a range of activities.

References

Put down the details (name, address and contact details) of two referees – usually a teacher and an employer, or somebody who knows you but is not related.

Worksheet 9

Layout of written information

Documents are set out in a particular way. This is referred to as a standard format. Depending upon the formality of a document the layout could be quite structured such as a letter or an agenda, or it could be quite flexible where the design is left up to the author.

Document formats can change by adjusting the font/size, adding graphics or colour.

1. Design a suitable document format for the following extracts, try to make them as eye catching as possible.

Document 1

A4 poster

Coming to a screen near you, 'Attack of the Droids'. Previews Thursday 2 May 2002. All seats £3.25. Aurora Cinemas.

Document 2

Business card

Contact Sensitive for all of your computing needs. David Hamble, 44 Conway Drive, Haddisham, Suffolk IP13 3DD, telephone 01463 0333330 e-mail: dhamble@contact.co.uk.

Document 3

Menu

Fursham's

Starters: watercress and wild mushroom tart, Scottish salmon terrine with dill and lemon sauce, roasted vegetable soup with herb croutons.

Champagne sorbet

Main course: free range chicken with apricot sauce, duck with ginger coulis, trout with almonds, Moroccan lamb, spinach and ricotta pancakes.

All served with green beans, shallots and chive and onion mash.

Dessert: pecan and honey tart, rich chocolate pudding with chestnut sauce, pistachio and cinnamon ice cream, kiwi and lemon sorbet.

Cheese and biscuits

Coffee and mints

1.4 Visual communication

Worksheet 10

Using visual formats

Visual communication includes a wide range of pictorial, graphical, design and interactive tools which enable a more effective, colourful understanding and conveyance of data and information.

1. There are a number of categories of visual communication. Identify six of these categories.
2. Represent the following information in a visual format:
    'From the bank turn right and walk to the first set of traffic lights (about 200 yards) when you get there turn left into Pine Street. Go past Acre Street on your left (50 yards) and Stone Walk on your left (100 yards). 50 yards past this on your right is the library on the corner of Lime Tree Street. Take the first right off Lime Tree Street into Market Square, the coffee shop is there on your left, opposite the book shop.'
3. The example of the draft design for a housing development indicates visually the plans of new houses to be built. Using a similar scale design a similar development to accommodate the following requirements:
    (i)   Twelve 3, seven 4 and six 5 bedroom houses
    (ii)   All of the 5 bedroom houses should have a double garage, three of the 4 bedroom houses should have a single garage and four of the 3 bedroom houses should have a single garage
    (iii) Swimming pool and two tennis courts
    (iv) Courtyard environment
    (v)   Good road links and pathways
    (vi) Lots of trees and foliage

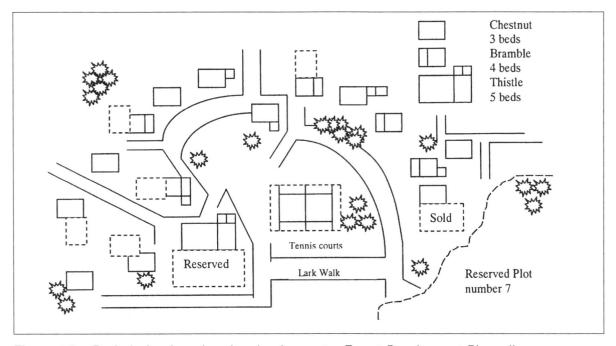

Figure 1.2 *Draft design for a housing development – Forest Development Phase II*

Worksheet 11

Advertisements

Moving images have become one of the most important forms of communication, especially with the ever-increasing popularity of television and movies. These formats of moving images have been used to advertise and sell, entertain and instruct and educate.

1. Advertisements for products and services are everywhere we go. Identify six different advertisement mediums.
2. For each of the different mediums provide a comparison in terms of:
   (i) Impact (how noticeable is it, is it eye catching, can you remember the advert days afterwards etc.?)
   (ii) Range of target audience (how many people would see the advert?)
   (iii) Limitations with the advertisement medium
3. Study TV adverts for a period of one week and produce a table identifying:
   (i) How many adverts were viewed
   (ii) Types of adverts – what were they for?
   (iii) Length of the advert
   (iv) Features of the advert – did it have music, a famous TV or film star, story-line or theme, animals etc.?
   (v) The orientation of the advert – was it funny, serious, formal etc.?
4. From the adverts viewed pick the best and the worst advert and describe why these were nominated under the two chosen categories.

Worksheet 12

Producing a television advert

You are a television advertisement design company and you have been approached by four clients who want you to design suitable adverts for their products.

- Client 1 needs a funny advert preferably starring someone from a soap series to advertise pet food (advert to last 1 minute).
- Client 2 requires a more mature TV personality to advertise garden fertilizer (advert to last 30 seconds).
- Client 3 has asked for a sports personality to advertise their new sporting holiday packages to include: sky diving, skiing and sailing.
- Client 4 wants an advert with a theme which can be used for adverts in the future, possibly a mini-drama. The product is deodorant.

1. Using a series of design boards create an advert for each client in accordance with their requirements.

For example:

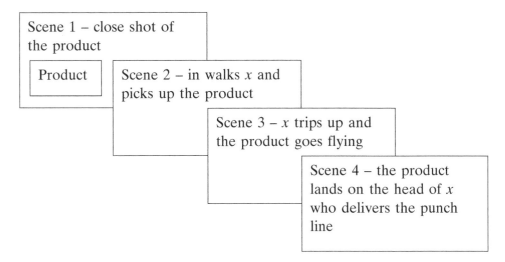

2. Each advert must have a punch line or slogan at the end.
3. In small groups of four each member should put together one of their adverts with the help of team members to act out in front of the class.

Worksheet 13

Using charts and graphs

Graphs and charts are used to provide visual support to data and tables providing a clear breakdown of key data components.

Using the information provided carry out the following tasks:

1. Using applications software type in the data given and produce a bar graph clearly labelling the data components.
2. Produce a pie chart identifying the market share of each car.
3. Produce a line graph to indicate the peaks and troughs of the market share over the period stated.
4. What trends or patterns have emerged from the graphs and charts produced.
5. Using the information gathered from task 4, was this made easier by looking at the graphs and charts or could this be seen easily from the information given in Table 1.1?

Table 1.1 *Market share % for car sales*

	Engine size	January	February	March	April	May	June
Car A	1.6	35	32	36	30	38	40
Car B	1.8	21	22	23	20	25	21
Car C	2.0	17	15	16	14	18	15
Car D	2.4	12	12	11	10	13	12
Car E	3.0	6	7	8	6	9	7

Worksheet 14

Expressive communication

Expressive communication is given to the category of communication that includes body signs, movements, gestures and language.

1. Identify six different forms of expressive communication.
2. For each of the forms listed, identify whether or not another form of either expressive or non-expressive communication could have been used.
3. For a period of 24 hours identify all of the different body language actions that you use and identify where, when and to whom you expressed these.
4. On reflection would it have been more appropriate to have used an alternative expressive action?
5. Did you use another form of communication in conjunction with the expressive action and why did you need to?

1.5   Presentations, discussions and debates

Worksheet 15

Presentations

Good presentation skills are essential for all walks of life and especially for progression into higher education, training or employment. Most people will be expected to give a presentation either informally at school or college as part of an assessment, or formally at work in front of colleagues, at a team meeting or to management personnel.

There is always something that you can do to improve your presentation skills, these tasks are designed to look at how you behave when you are giving a presentation and hopefully can be used to improve your presentation skills.

1. In pairs assume the role of reviewers taking it in turn to talk for five minutes about a recent film that you have watched. In preparation for this write down a list of points about:
    (i)   Who the main characters were
    (ii) The plot and/or subplots
    (iii) The graphics and special effects
    (iv) The costumes
    (v)   The realism
2. As your partner is talking make comments on certain aspects of their presentation – complete a feedback sheet as shown in the example.
3. Following the presentations, each person should feed back to the other.
4. Deliver the review again and try to address some of the feedback given to improve and enhance the quality of your presentation.

Feedback template

Name: Date:

Presentation type: Formal ☐   Informal ☐

Presentation Title:

Checklist

	Yes:	No:
Introduction given:	☐	☐
Clear and concise throughout:	☐	☐
Presentation closed:	☐	☐
Variation in voice tone:	☐	☐
Acknowledges audience:	☐	☐
Establishes eye contact:	☐	☐
Good pace of presentation:	☐	☐

Comments regarding delivery:

Comments regarding body language:

Comments regarding tools used (if applicable):

Areas for improvement:
1.
2.
3.

Strengths
1.
2.
3.

Worksheet 16

Formal presentations

You have been asked to give a formal presentation which has been well researched about one of the following topic areas:

- 'Mobile phones – do they make you feel more secure because of instant communication or do they make you a target for crime?'
- 'The internet – friend or foe?'
- 'Online shopping – do we really need to support high street stores?'

1. You have one week to research one of the three topic areas and give a 15 minute presentation on your findings.
2. Each topic researched should provide evidence in support and against the statements given.
3. The presentations should last no more than fifteen minutes with time at the end for questions.
4. Your presentation should try to incorporate at least one recognized tool, e.g. an OHP, video or the use of presentation software.

Worksheet 17

Freestyle presentation

It is sometimes easier to talk about a topic that you are familiar with, and do not need to research or remember facts and figures.

1. Prepare a 20 minute presentation to be delivered in front of the class. The presentation should be based on a topic that you are interested in or a hobby or pastime, e.g.:
   (i) A team that you support
   (ii) A particular sporting activity
   (iii) Your involvement with a club or society
   (iv) A collection of memorabilia
   (v) A hobby, e.g. playing an instrument or reading etc.
   (vi) Favourite television programme, film, group etc.
2. For the presentation try to bring in items to support what you are doing, e.g. certificates, sports shirts, video footage, pictures, clippings, sound clips etc.

Worksheet 18

Debates

Debating is a very important skill which takes self-control and discipline, especially if you are debating a topic which is close to your heart.

1. The class should be divided in two even numbered groups each representing a debate group. Within each debate group a further equal division should be made for those in support of an issue and those against.

2. Each debate group should select a topic from the ones listed, or choose their own topic, topics available include:
   (i) Is there life on other planets?
   (ii) Does the Loch Ness monster really exist?
   (iii) Should cannabis be legalized?
   (iv) Should capital punishment be brought back?
   (v) Should euthanasia be made legal in the United Kingdom?
3. Each debate group should then decide who is going to argue in favour and who is going to argue against.
4. All arguments should be supported by evidence and research rather than personal opinions.
5. At the end of the debate the other group should state who had the more convincing argument and why.

1.6 Read and respond to written material

Worksheet 19

Why and how we read

The ability to read and absorb information can vary to a degree depending upon a number of factors.

1. Describe how and why the following factors could influence your ability to read and absorb information:
   (i) The subject matter
   (ii) Level of material
   (iii) Presentation of material
2. Provide an example of how each of the factors in task 1 has affected how you read and have absorbed information in the past.
3. The more interesting the text, the higher the chances of understanding the theme which will improve the chances of completion. Do you agree or disagree with this statement?
4. Identify three reasons why you read.
5. State two reasons for reading the following texts:
   (i) National Diploma course textbook
   (ii) Computing magazine
   (iii) Newspaper
   (iv) Novel
   (v) Software manual

1.7 Reading and note taking

Worksheet 20

Note taking methods

People take notes every day to jot down facts, summarize information such as a lesson or to record data such as times or activities. The way in which people take notes can depend on how an individual interprets the information, and how they are able to process and reflect what is being stated or read into a written or visual format.

1. Read the following extract and convert it into note format using:
    (i)   Mind map
    (ii) Bullet points

    'Communication is a way of expressing thoughts and ideas from one person or party to another. To enable effective communication there needs to be three main elements: a sender, a communication tool and a receiver. The sender of information is sometimes referred to as the source of information. All information will originate from a source of some description. The source generates or initiates the information that is to be sent. Sources of information can include human resources (people) and electronic resources (computer systems etc.), these resources are also recognized as receivers of information. The type of communication tool that is required to transmit the information will vary depending upon a number of factors, these factors include: time, impact, expense, geography, convenience, audience, occasion and formality.'

2. Identify three other ways that you could generate notes from the extract.
3. Which note taking method would be easiest to:
    (i)   Produce?
    (ii) Understand?
    (iii) Relay to another person?

1.8 Punctuation, spelling and grammar

Worksheet 21

Punctuation

1. Complete the table by filling in the missing information about punctuation symbols.

Punctuation provides the clarity, breaks and context to a sentence. Punctuation allows you to break down written extracts into manageable chunks through the use of:

Syntax	Descriptor	Some examples of use
'	Apostrophe	Can state possession if used with the letter 's', e.g. the student's work
:		Indicates stops or breaks in clauses, usually used to introduce examples
		Used to indicate breaks in a sentence between groups of words or a list of items
–		Denotes additionality to a sentence
		Specifies what was spoken and by whom, also identifies a quotation or title
	Exclamation mark	Denotes the feeling of surprise or other emotive expressions
		Represents the end of a sentence or an abbreviated word, for example e.g. or etc.
		Link between words or parts of words
	Parentheses	Separate primary and secondary ideas
		Represents the ending of a direct question
	Semicolon	A pause which is longer than a comma but shorter than a full-stop, a dramatic pause

Worksheet 22

Words and grammar

The following are words that sound the same but are spelt differently and are known as homophones:

fair	fare
their	there
horse	hoarse
meet	meat
flee	flea
compliment	complement
past	passed
claws	clause

1. Provide six more examples of homophones.
2. What do the terms 'prefix' and 'suffix' mean?
3. Using the example below provide five more examples of prefix and suffix words:

e.g.   disappear ⟶ appear ⟶ disappearing

          prefix   suffix

4. For each of the following words provide a plural alternative.

Word	Plural alternative
Copy	
Chase	
Person	
Wave	
Happy	
Tomato	
Usual	
Worry	
Shelf	
Sheep	
Lunch	
Dice	
Swim	
Piece	
Fish	

5. Complete the information in the following grammar table.

Identifier	Description	Example
Adjective	Describing word	large, pretty, heavy, cold
Adverb	Extension to adjectives or verbs: how, where, when	
	Definite and indefinite	*the* assignment *a* job
Conjunction		and, next, when, if, after
Interjection		ouch, yippee, oh, wow
	Naming word	
Preposition	Locators	
	Naming or identifying words	refers to people or things: *Who* did you invite?
Verb		

Worksheet 23

Spelling

1. For each of the following sets of words identify the correct spelling **do not use spell check**.

Commplient	Complament	Compliment	Complimant
Existance	Existence	Existense	Existanse
Friendly	Freindly	Friendely	Friendally
Satisfyed	Satisfied	Satissfied	Satesfied
Compleete	Compleat	Complete	Compleit
Asessment	Assesment	Assessment	Asesment
Fortunate	Fortunute	Fortunet	Fortunat
Thorough	Thorogh	Thorugh	Thorrow
Orderlay	Orderley	Orderly	Ordarly
Bording	Boarding	Boreding	Boarrding
Striving	Streiving	Strieving	Stryving
Discused	Discussed	Disscused	Disscussed
Planit	Plannet	Planat	Planet
Becouse	Becase	Because	Becauce
Achieved	Acheived	Acheeved	Acheaved

Worksheet 24

Using correct spelling, grammar and punctuation

Eight-Ball Gadgets and Gizmos

Sales Assistant
£4.50 an hour + commission

We are looking to recruit a new part-time member to our sales team for our branch in Norwich. Must be able to work evenings and weekends for a maximum of 15 hours. Previous shop experience is desirable and a passion for all gadgets and gizmos is essential.

For further information e-mail: sblackwell@eightball.co.uk
To apply forward your CV to:

Simon Blackwell
34 The Arcade
Norwich
NR15 6DR
Tel: 01603 5778990

Closing date: 10 May 2002

---

1. Write an e-mail to Simon Blackwell asking for more details about the job, hours and duties and send this to a friend within the group. Do not adjust the spelling if prompted by a spell check.
2. Together go through each e-mail and comment on the content, grammar, punctuation and spelling.
3. Discuss any problems and either check back with your tutor or use the correction functions of the e-mail or word processing package to check your concerns on your list.
4. As a group discuss what grammatical errors were made to see if there is any pattern in regards to difficult words or sentences.

Worksheet 25

Identifying errors

1. Examine the extract and identify the punctuation, spelling and grammatical errors.

'your advertisement in Buzz Computing on Friday 17 invited applications from keen gaming players to subscribe to a new computer interactive group being set up in East Anglia.

My interests in computers and games consoles has incresed over the last year with me acquiring a number of rare games consoles and software, which can be considered as "retro" items. i also enjoy interactive on-line adventure gaming sessions and role play scenarios.

I feel that I could contributed to your group in a number of ways. Firstly I have an extensive knowledge of hardware and software. I am also a keen programer and a software enthusiast, I also submit regular on-line reviews of new games to the "What's New in Computing" magazine.

I enclose a check for £85.00 in respect of twelve months subscription to the your computing group, and I hope that you accept my aplication.

I look forward to hearing from you in the near future.'

2. Which errors did you find more difficult to recognize. Why do you think this was?

1.9 Research skills

Worksheet 26

Carrying out research

To be an effective communicator includes listening, reading and writing skills. Arguably another skill that is required to complete the package is that of research.

1. Complete the table by stating the advantages and disadvantages of each research tool.

Research tool	Advantages	Disadvantages
Book		
Internet		
People		
TV/Radio		
Magazine		

2. Using two of the research tools listed in the table, carry out research and prepare an A4 sheet on one of the following topic areas:
   (i)   The benefits of e-commerce (buying and trading online)
   (ii) Footballers – are they worth what they are paid?
   (iii) Third world debt – are we as a country doing enough?
   (iv) The rise in student debt – is this deterring more youngsters from going onto college or university?

Worksheet 27

Preparing a research plan

Using a research plan to co-ordinate your work and provide structure to your research is a useful aid.

There are a number of different ways of recording information and resources, quite simply preparing a list or taking down appropriate reference numbers or e-mail addresses. However, to ensure that resources are recorded accurately, possibly for future use, it is better to use a resource template as shown in the examples.

1. Design a suitable resource template plan which can be used as a standard template for recording either one or multiple resources for future assignments.
2. For an assignment that you have been given use the appropriate research template to document the resources to be used.
3. Identify four benefits of using a research plan.
4. When you have used one of your research plans comment on the following:
   (i)   Did it help enhance the quality of your assignment?
   (ii) Did you select more resources as a result of using the plan?
   (iii) Did you use a wider variety of resources?

Internet site research template

Search engine: (if applicable)	Web page address:
Website category: (academic, commercial, industrial)	Links used:
Summary of site information:	
Book marked: Yes ☐ No ☐	

Book research template

Book title	Author
Date of publication:	Publisher:
Edition:	ISBN:
Page/s viewed:	
Information summary:	

General research plan template

Research topic	
Start date:	Completion date:
Objectives of the research Tasks: 1. 2. 3. 4. 5.	
Task dependencies: (are any of the tasks reliant upon another etc.)	
Resources to be used:	Location and reference of resources:
Allocation of time:	
Additional information:	

1.10   Barriers to communication

Worksheet 28

Acknowledging communication barriers

There are many barriers that can constrain people from communicating with each other.

1. Barriers to communication can fall under a number of categories, for example:
    (i)   Geographical
    (ii) Social
    (iii) Cultural
    (iv) Technological
    For each category listed provide an example of how each can cause a barrier to communication.
2. What methods could you use to communicate with somebody who did not speak your language?
3. Customs vary from country to country and what is seen as a positive and friendly gesture in one culture can be considered an offence in another. Identify four countries' customs and describe what the impact is upon that country and also for people who want to visit that country (what would they have to do) and what might happen if they didn't?

Unit 2 Computer systems

2.1 Stored data

Integers

A file of Pascal data, 12 bytes long, is known to contain six numbers, the first number has the value 34. What are the other numbers in the file? Attempts to show these numbers directly on the screen using the Windows Notepad program yielded the result:

" ‰ [
íu¡Úÿÿ

which are clearly not numbers if interpreted as ASCII!

In a further attempt to discover the values of these numbers, the DOS DEBUG program was used to provide a hex dump of the file. This yielded the following:

```
0E72:0100   22 00 89 00 5B 0D ED 75-A1 DA FF FF 0D 74 02 EB   "...[..u.....t..
0E72:0110   F8 8B CF 81 E9 82 00 26-88 0E 80 00 34 00 61 0E   .......&....4.a.
0E72:0120   DE BE 10 D4 BA FF FF B8-00 AE CD 2F 3C 00 C3 A0   ........../<...
0E72:0130   D1 E2 0A C0 74 09 56 57-E8 2A 21 5F 5E 73 0A B9   ....t.VW.*!_^s..
0E72:0140   04 01 FC 56 57 F3 A4 5F-5E C3 50 56 33 C9 33 DB   ...VW.._^.PV3.3.
0E72:0150   AC E8 5F 23 74 19 3C 0D-74 15 F6 C7 20 75 06 3A   .._#t.<.t... u.:
0E72:0160   06 02 D3 74 0A 41 3C 22-75 E6 80 F7 20 EB E1 5E   ...t.A<"u... ..^
0E72:0170   58 C3 A1 D7 D7 8B 36 D9-D7 C6 06 1B D9 00 C6 06   X.....6.........
```

(See Appendix B for details of the DEBUG program.)

As the file size was just 12 bytes, it seems obvious that each number is stored in just 2 bytes. The first 12 bytes split into six pairs are:

22	00	89	00	5B	0D	ED	75	A1	DA	FF	FF

Question

What are the other numbers in the file?

Answer

The first number has the value 34 which is 22 in hex. The byte pair 22 00 seems to agree but is the 'wrong way round'. It would seem a reasonable assumption that the numbers are 16-bit numbers stored low byte:high byte.

Putting the values 'the other way' round yields:

0022
0089
0D5B
75ED
DAA1
FFFF

If these values are now converted to binary we get:

Hex	Binary
0022	0000000000100010
0089	0000000010001001
0D5B	0000110101011011
75ED	0111010111101101
DAA1	1101101010100001
FFFF	1111111111111111

There is now a problem. The left-hand bits of the last two of these numbers is set to 1. There is no way to tell by inspection if the numbers are signed integers, i.e. if this left-hand bit is used as a sign bit or if they are 16-bit unsigned numbers, so the best that is possible is to give both possibilities as shown in the table.

Hex	Binary	Unsigned integer	Signed integer
0022	0000000000100010	34	34
0089	0000000010001001	137	137
0D5B	0000110101011011	3419	3419
75ED	0111010111101101	30189	30189
DAA1	1101101010100001	55969	-9567
FFFF	1111111111111111	65535	-1

Explain the original screen output from Notepad

The result was:

" ‰ [
íu¡Úÿÿ

The file has the 12 bytes:

22 00 89 00 5B 0D ED 75 A1 DA FF FF

If these are shown as ASCII characters, all is clear. (See Appendix A for ASCII table.) The file contains hex byte 0D which is a Carriage Return, hence the break in the data after the [character.

Hex	22	00	89	00	5B	0D	ED	75	A1	DA	FF	FF
ASCII	"	NUL	‰	NUL	[	CR	í	u	¡	Ú	ÿ	ÿ

By looking at just a hex dump, there is no way to prove exactly what the data is. The numbers can be decoded as integers provided that it is known the data is made up of integers; the data could well be a set of values in a completely different format. In this case, it is unlikely the data is ASCII as many ASCII files are 'human readable', i.e. you can make out what is in the file by inspection, even if the file contains some control characters.

The only way that data can be assigned real meaning is to open the data in the software used to produce the file or software that 'knows' what the data is by some other means.

As a further example, you may have recorded the following numerical measurement data in decimal and stored it as unsigned 8-bit bytes:

74	97	109	101	115	32	98	111	110	100

You would then be surprised to find that if you use a piece of software like Word or Notepad, your set of numbers seem to spell a well-known name! You either decode this by hand or use one of the ASCII functions of a spreadsheet.

Little endian numbers

Some systems store 16-bit values as low byte:high byte but others store them as high byte:low byte. When numbers are stored low byte:high byte it is known as 'little endian', i.e. the little number is stored first (or at the lowest address in memory). Systems that store numbers the other way round are known as 'big endian'. This all started in the design of microprocessors so was fixed in the circuitry. Many modern systems are able to deal with both ways of storage.

Worksheet 1

Stored numbers

Six numbers are stored in the first 12 bytes of a file. It is known that they are 16-bit little endian values but it is not known if they are signed or unsigned.

Write down all six numbers as both signed and unsigned decimal values.

```
0E73:0100   B7 00 59 00 68 5D A3 FE-09 25 19 E3 C3 8B 1E 88   ..Y.h]...%......
0E73:0110   DE BE 10 D4 BA FF FF B8-00 AE CD 2F 34 00 62 0E   ........../4.b.
0E73:0120   D1 E2 0A C0 74 09 56 57-E8 2A 21 5F 5E 73 0A B9   ....t.VW.*!_^s..
0E73:0130   04 01 FC 56 57 F3 A4 5F-5E C3 50 56 33 C9 33 DB   ...VW.._^.PV3.3.
0E73:0140   AC E8 5F 23 74 19 3C 0D-74 15 F6 C7 20 75 06 3A   .._#t.<.t... u.:
0E73:0150   06 02 D3 74 0A 41 3C 22-75 E6 80 F7 20 EB E1 5E   ...t.A<"u... ..^
0E73:0160   58 C3 A1 D7 D7 8B 36 D9-D7 C6 06 1B D9 00 C6 06   X.....6.........
0E73:0170   17 D9 00 8B 36 D9 D7 8B-0E D7 D7 8B D6 E3 42 51   ....6.........BQ
```

Signed	Unsigned

Worksheet 2

Stored numbers

Six numbers are stored in the first 12 bytes of a file. It is known that they are 16-bit little endian values but it is not known if they are signed or unsigned.

Write down all six numbers as both signed and unsigned decimal values.

```
0E72:0100   86 6F BE 72 D4 3B 03 72-44 68 65 10 0D 74 02 EB   .o.r.;.rDhe..t..
0E72:0110   F8 8B CF 81 E9 82 00 26-88 0E 80 00 34 00 61 0E   .......&....4.a.
0E72:0120   DE BE 10 D4 BA FF FF B8-00 AE CD 2F 3C 00 C3 A0   ........../<...
0E72:0130   D1 E2 0A C0 74 09 56 57-E8 2A 21 5F 5E 73 0A B9   ....t.VW.*!_^s..
0E72:0140   04 01 FC 56 57 F3 A4 5F-5E C3 50 56 33 C9 33 DB   ...VW.._^.PV3.3.
0E72:0150   AC E8 5F 23 74 19 3C 0D-74 15 F6 C7 20 75 06 3A   .._#t.<.t... u.:
0E72:0160   06 02 D3 74 0A 41 3C 22-75 E6 80 F7 20 EB E1 5E   ...t.A<"u... ..^
0E72:0170   58 C3 A1 D7 D7 8B 36 D9-D7 C6 06 1B D9 00 C6 06   X.....6.........
```

Signed	Unsigned

Worksheet 3

Stored numbers

Six numbers are stored in the first 12 bytes of a file. It is known that they are 16-bit little endian values but it is not known if they are signed or unsigned.

Write down all six numbers as both signed and unsigned decimal values.

```
0E72:0100   00 16 CE 37 80 29 18 22-DE 32 91 65 0D 74 02 EB   ...7.)."..2.e.t..
0E72:0110   F8 8B CF 81 E9 82 00 26-88 0E 80 00 34 00 61 0E   .......&....4.a.
0E72:0120   DE BE 10 D4 BA FF FF B8-00 AE CD 2F 3C 00 C3 A0   .........../<...
0E72:0130   D1 E2 0A C0 74 09 56 57-E8 2A 21 5F 5E 73 0A B9   ....t.VW.*!_^s..
0E72:0140   04 01 FC 56 57 F3 A4 5F-5E C3 50 56 33 C9 33 DB   ...VW.._^.PV3.3.
0E72:0150   AC E8 5F 23 74 19 3C 0D-74 15 F6 C7 20 75 06 3A   .._#t.<.t... u.:
0E72:0160   06 02 D3 74 0A 41 3C 22-75 E6 80 F7 20 EB E1 5E   ...t.A<"u... ..^
0E72:0170   58 C3 A1 D7 D7 8B 36 D9-D7 C6 06 1B D9 00 C6 06   X.....6.........
```

Signed	Unsigned

Worksheet 4

Stored numbers

Six numbers are stored in the first 12 bytes of a file. It is known that they are 16-bit little endian values but it is not known if they are signed or unsigned.

Write down all six numbers as both signed and unsigned decimal values.

```
0E72:0100   4C 70 4A 2B BC 65 34 49-6A 66 1D 22 0D 74 02 EB   LpJ+.e4Ijf.".t..
0E72:0110   F8 8B CF 81 E9 82 00 26-88 0E 80 00 34 00 61 0E   .......&....4.a.
0E72:0120   DE BE 10 D4 BA FF FF B8-00 AE CD 2F 3C 00 C3 A0   .........../<...
0E72:0130   D1 E2 0A C0 74 09 56 57-E8 2A 21 5F 5E 73 0A B9   ....t.VW.*!_^s..
0E72:0140   04 01 FC 56 57 F3 A4 5F-5E C3 50 56 33 C9 33 DB   ...VW.._^.PV3.3.
0E72:0150   AC E8 5F 23 74 19 3C 0D-74 15 F6 C7 20 75 06 3A   .._#t.<.t... u.:
0E72:0160   06 02 D3 74 0A 41 3C 22-75 E6 80 F7 20 EB E1 5E   ...t.A<"u... ..^
0E72:0170   58 C3 A1 D7 D7 8B 36 D9-D7 C6 06 1B D9 00 C6 06   X.....6.........
```

Signed	Unsigned

Worksheet 5

Stored numbers

Six numbers are stored in the first 12 bytes of a file. It is known that they are 16-bit little endian values but it is not known if they are signed or unsigned.

Write down all six numbers as both signed and unsigned decimal values.

```
0E72:0100   17 59 AE 64 36 28 6D 1E-30 0E 21 70 0D 74 02 EB   .Y.d6(m.0.!p.t..
0E72:0110   F8 8B CF 81 E9 82 00 26-88 0E 80 00 34 00 61 0E   .......&....4.a.
0E72:0120   DE BE 10 D4 BA FF FF B8-00 AE CD 2F 3C 00 C3 A0   .........../<...
0E72:0130   D1 E2 0A C0 74 09 56 57-E8 2A 21 5F 5E 73 0A B9   ....t.VW.*!_^s..
0E72:0140   04 01 FC 56 57 F3 A4 5F-5E C3 50 56 33 C9 33 DB   ...VW.._^.PV3.3.
0E72:0150   AC E8 5F 23 74 19 3C 0D-74 15 F6 C7 20 75 06 3A   .._#t.<.t... u.:
0E72:0160   06 02 D3 74 0A 41 3C 22-75 E6 80 F7 20 EB E1 5E   ...t.A<"u... ..^
0E72:0170   58 C3 A1 D7 D7 8B 36 D9-D7 C6 06 1B D9 00 C6 06   X.....6.........
```

Signed	Unsigned

Worksheet 6

Stored numbers

Six numbers are stored in the first 12 bytes of a file. It is known that they are 16-bit little endian values but it is not known if they are signed or unsigned.

Write down all six numbers as both signed and unsigned decimal values.

```
0E72:0100   C7 8D C7 9C 73 3B 30 3C-36 FD 8A CF 0D 74 02 EB   ....s;0<6....t..
0E72:0110   F8 8B CF 81 E9 82 00 26-88 0E 80 00 34 00 61 0E   .......&....4.a.
0E72:0120   DE BE 10 D4 BA FF FF B8-00 AE CD 2F 3C 00 C3 A0   .........../<...
0E72:0130   D1 E2 0A C0 74 09 56 57-E8 2A 21 5F 5E 73 0A B9   ....t.VW.*!_^s..
0E72:0140   04 01 FC 56 57 F3 A4 5F-5E C3 50 56 33 C9 33 DB   ...VW.._^.PV3.3.
0E72:0150   AC E8 5F 23 74 19 3C 0D-74 15 F6 C7 20 75 06 3A   .._#t.<.t... u.:
0E72:0160   06 02 D3 74 0A 41 3C 22-75 E6 80 F7 20 EB E1 5E   ...t.A<"u... ..^
0E72:0170   58 C3 A1 D7 D7 8B 36 D9-D7 C6 06 1B D9 00 C6 06   X.....6.........
```

Signed	Unsigned

Worksheet 7

Stored numbers

Six numbers are stored in the first 12 bytes of a file. It is known that they are 16-bit little endian values but it is not known if they are signed or unsigned.

Write down all 6 numbers as both signed and unsigned decimal values.

```
0E72:0100   EA CB 25 1A 8C 07 9D C6-DD 4C 1F BD 0D 74 02 EB   ..%......L...t..
0E72:0110   F8 8B CF 81 E9 82 00 26-88 0E 80 00 34 00 61 0E   .......&....4.a.
0E72:0120   DE BE 10 D4 BA FF FF B8-00 AE CD 2F 3C 00 C3 A0   ........../<...
0E72:0130   D1 E2 0A C0 74 09 56 57-E8 2A 21 5F 5E 73 0A B9   ....t.VW.*!_^s..
0E72:0140   04 01 FC 56 57 F3 A4 5F-5E C3 50 56 33 C9 33 DB   ...VW.._^.PV3.3.
0E72:0150   AC E8 5F 23 74 19 3C 0D-74 15 F6 C7 20 75 06 3A   .._#t.<.t... u.:
0E72:0160   06 02 D3 74 0A 41 3C 22-75 E6 80 F7 20 EB E1 5E   ...t.A<"u... ..^
0E72:0170   58 C3 A1 D7 D7 8B 36 D9-D7 C6 06 1B D9 00 C6 06   X.....6.........
```

Signed	Unsigned

Worksheet 8

Stored numbers

Six numbers are stored in the first 12 bytes of a file. It is known that they are 16-bit little endian values but it is not known if they are signed or unsigned.

Write down all six numbers as both signed and unsigned decimal values.

```
0E72:0100   8C 71 58 4A 8C 52 A3 8B-F8 5B 0D C6 0D 74 02 EB   .qXJ.R...[...t..
0E72:0110   F8 8B CF 81 E9 82 00 26-88 0E 80 00 34 00 61 0E   .......&....4.a.
0E72:0120   DE BE 10 D4 BA FF FF B8-00 AE CD 2F 3C 00 C3 A0   ........../<...
0E72:0130   D1 E2 0A C0 74 09 56 57-E8 2A 21 5F 5E 73 0A B9   ....t.VW.*!_^s..
0E72:0140   04 01 FC 56 57 F3 A4 5F-5E C3 50 56 33 C9 33 DB   ...VW.._^.PV3.3.
0E72:0150   AC E8 5F 23 74 19 3C 0D-74 15 F6 C7 20 75 06 3A   .._#t.<.t... u.:
0E72:0160   06 02 D3 74 0A 41 3C 22-75 E6 80 F7 20 EB E1 5E   ...t.A<"u... ..^
0E72:0170   58 C3 A1 D7 D7 8B 36 D9-D7 C6 06 1B D9 00 C6 06   X.....6.........
```

Signed	Unsigned

Worksheet 9

Stored numbers

Six numbers are stored in the first 12 bytes of a file. It is known that they are 16-bit little endian values but it is not known if they are signed or unsigned.

Write down all six numbers as both signed and unsigned decimal values.

```
0E72:0100   6B  28  FF  64  2B  13  1C  E4-A9  BA  2A  00  0D  74  02  EB   k(.d+.....*..t..
0E72:0110   F8  8B  CF  81  E9  82  00  26-88  0E  80  00  34  00  61  0E   .......&....4.a.
0E72:0120   DE  BE  10  D4  BA  FF  FF  B8-00  AE  CD  2F  3C  00  C3  A0   .........../<...
0E72:0130   D1  E2  0A  C0  74  09  56  57-E8  2A  21  5F  5E  73  0A  B9   ....t.VW.*!_^s..
0E72:0140   04  01  FC  56  57  F3  A4  5F-5E  C3  50  56  33  C9  33  DB   ...VW.._^.PV3.3.
0E72:0150   AC  E8  5F  23  74  19  3C  0D-74  15  F6  C7  20  75  06  3A   .._#t.<.t... u.:
0E72:0160   06  02  D3  74  0A  41  3C  22-75  E6  80  F7  20  EB  E1  5E   ...t.A<"u... ..^
0E72:0170   58  C3  A1  D7  D7  8B  36  D9-D7  C6  06  1B  D9  00  C6  06   X.....6.........
```

Signed	Unsigned

Worksheet 10

Stored numbers

Six numbers are stored in the first 12 bytes of a file. It is known that they are 16-bit little endian values but it is not known if they are signed or unsigned.

Write down all six numbers as both signed and unsigned decimal values.

```
0E72:0100   80  8C  03  5F  FB  16  1F  07-12  6E  AC  02  0D  74  02  EB   ..._.....n...t..
0E72:0110   F8  8B  CF  81  E9  82  00  26-88  0E  80  00  34  00  61  0E   .......&....4.a.
0E72:0120   DE  BE  10  D4  BA  FF  FF  B8-00  AE  CD  2F  3C  00  C3  A0   .........../<...
0E72:0130   D1  E2  0A  C0  74  09  56  57-E8  2A  21  5F  5E  73  0A  B9   ....t.VW.*!_^s..
0E72:0140   04  01  FC  56  57  F3  A4  5F-5E  C3  50  56  33  C9  33  DB   ...VW.._^.PV3.3.
0E72:0150   AC  E8  5F  23  74  19  3C  0D-74  15  F6  C7  20  75  06  3A   .._#t.<.t... u.:
0E72:0160   06  02  D3  74  0A  41  3C  22-75  E6  80  F7  20  EB  E1  5E   ...t.A<"u... ..^
0E72:0170   58  C3  A1  D7  D7  8B  36  D9-D7  C6  06  1B  D9  00  C6  06   X.....6.........
```

Signed	Unsigned

2.2 Bitmapped graphics files

The BTEC unit specifies that students should understand the basic concepts and principles involved in the representation of data including bitmaps. A single bitmapped file is presented here to show the contents and file structure.

The picture in Figure 2.1 is stored in a common file format called a 'bitmap'. (Under the Windows operating system, such files will have the file extension .BMP.)

Figure 2.1 *Single bitmapped file*

The file is 54 by 76 pixels and was drawn with 256 colours. This means that each pixel is stored in 1 byte in the file so there are $54 \times 76 = 4104$ pixels in the whole image. The file is 5334 bytes in size, the rest of the 5334-4104 bytes of data are stored in a 'file header' which contains mainly a 'colour palette', a means to specify which colours are to be used.

A typical system will use 3 bytes to store each pixel, one for the red intensity, one for green and one for blue. Any of 16.7 million colours can then be made as 3 bytes = 24 bits and $2^{24} = 16.7$ million.

In an image like this one that can store only 256 colours per pixel, each pixel is represented by 1 byte and the 'palette' holds 256 entries, and each entry holds a 3-byte colour. For example, to find the actual colour of 'colour number' 38 in this file, you would look at entry number 38 in the palette, there you would find 3 bytes, one for red, one for green and one for blue, the actual colour of the pixel. Although the colour of any pixel can be one of 16.7 million colours, in a 256 colour file like this one, you can only have 256 choices from the 16.7 million colours.

The maximum number of colours in a bitmapped file are shown in Table 2.1.

Table 2.1 *Maximum number of colours in a bitmapped file*

Bits per pixel	Maximum number of colours	Known as
2	$2^2 = 4$	
4	$2^4 = 16$	
8	$2^8 = 256$	
16	$2^{16} = 65536$	Hi-Color
24	$2^{24} = 16777216$	Tru-Color or 24-bit colour

The file header is in two parts, the *bitmap file header* and the *bitmap information header*. These give information such as image width and height, number of colours per pixel etc. as shown in Table 2.2. The 'start position' refers to the byte position in the file so, for example, if you look 19 bytes into the file you will find the width of the image in pixels, shown in hex, stored *little endian*, i.e. with the least significant byte (or little) byte stored first.

This image is 54 pixels wide which is 36 hex. Starting from byte position 19, the file contains the 4 bytes 36 00 00 00. Because this is little endian, turn it round to *big endian format* to give 00 00 00 36, the image width. If the byte sequence were 23 A4 3D 00, the file size would be 3DA423 or 4 039 715 bytes.

In a similar way, looking at byte position 29, you get the 2 bytes 08 00, turned round this gives the value 8 bits or 1 byte per pixel, correct for a 256 colour image. At byte position 11 you get 36 04 00 00, turned round to big endian format this gives 436 (hex) or 1078. This is the offset to the bitmap data, i.e. the image data starts 1078 bytes into the file. Oddly, the image data is stored back to front, the last row of data is the first row of pixels in the image.

Table 2.2 *Bitmap image file format details*

The bitmap file header

Start position	Size in bytes	Meaning or use
1	2	set to 'BM' to show this is a .bmp file.
3	4	size of the file in bytes.
7	2	always set to zero.
9	2	always set to zero.
11	4	offset from the beginning of the file to the bitmap data.

The bitmap information header

15	4	size of the bitmap information header structure, in bytes.
19	4	width of the image, in pixels.
23	4	height of the image, in pixels.
27	2	set to zero.
29	2	number of bits per pixel.
31	4	compression, usually set to zero, 0 = no compression.
35	4	size of the image data, in bytes.
39	4	horizontal pixels per metre, usually set to zero.
43	4	vertical pixels per metre, usually set to zero.
47	4	number of colours used in the bitmap.
51	4	number of colours that are 'important' for the bitmap; if set to zero, all colours are important.

The bit map data

The data in Table 2.3 has a format that depends on the information in the header. If the file is a 24-bit file (16.7 million colours), there are 3 bytes per pixel, one for red intensity, one for green and one for blue. There is no palette.

If the file is 1 byte per pixel (256 colours), each pixel is just one value that is used with the palette to determine the colour of the pixel.

Table 2.3 *Hex dump of image in Figure 2.1*

*File position of
first item in row File contents (hex)*

1	42	4D	D6	14	00	00	00	00-00	00	36	04	00	00	28	00
17	00	00	36	00	00	00	4C	00-00	00	01	00	08	00	00	00
33	00	00	A0	10	00	00	00	0B-00	00	00	0B	00	00	00	01
49	00	00	00	01	00	00	10	08-10	00	18	10	18	00	21	18
65	18	00	21	21	18	00	29	21-18	00	29	29	18	00	21	18
81	21	00	29	21	21	00	31	21-21	00	29	29	21	00	31	29
97	21	00	39	29	21	00	31	31-21	00	39	31	21	00	39	39
113	21	00	31	29	29	00	39	29-29	00	31	31	29	00	39	31
129	29	00	42	31	29	00	39	39-29	00	42	39	29	00	4A	42
145	29	00	31	31	31	00	39	31-31	00	42	31	31	00	39	39
161	31	00	42	39	31	00	4A	39-31	00	42	42	31	00	4A	42
177	31	00	4A	4A	31	00	52	4A-31	00	31	31	39	00	42	39
193	39	00	4A	39	39	00	42	42-39	00	4A	42	39	00	52	42
209	39	00	52	4A	39	00	5A	52-39	00	4A	42	42	00	52	42
225	42	00	4A	4A	42	00	52	4A-42	00	52	52	42	00	5A	52
241	42	00	63	5A	42	00	42	42-4A	00	4A	42	4A	00	52	42
257	4A	00	4A	4A	4A	00	52	4A-4A	00	5A	4A	4A	00	52	52
273	4A	00	5A	52	4A	00	5A	5A-4A	00	63	5A	4A	00	5A	52
289	52	00	63	52	52	00	5A	5A-52	00	63	5A	52	00	63	63
305	52	00	52	52	5A	00	5A	52-5A	00	5A	5A	5A	00	63	5A
321	5A	00	6B	5A	5A	00	63	63-5A	00	6B	63	5A	00	6B	6B
337	5A	00	73	6B	5A	00	5A	52-63	00	63	5A	63	00	63	63
353	63	00	6B	63	63	00	6B	6B-63	00	73	6B	63	00	73	73
369	63	00	63	63	6B	00	6B	63-6B	00	6B	6B	6B	00	73	6B
385	6B	00	73	73	6B	00	7B	73-6B	00	84	7B	6B	00	6B	63
	Some	data	deleted	here	to	save	space								
1009	FF	00	FF	FF	FF	00	FF	FF-FF	00	FF	FF	FF	00	FF	FF
1025	FF	00	FF	FF	FF	00	FF	FF-FF	00	FF	FF	FF	00	FF	FF
1041	FF	00	FF	FF	FF	00	FF	FF-FF	00	FF	FF	FF	00	FF	FF
1057	FF	00	FF	FF	FF	00	FF	FF-FF	00	FF	FF	FF	00	FF	FF
1073	FF	00	FF	FF	FF	00	BB	BB-BB	BB	BB	BB	BB	BB	BB	BB
1089	BB	BB	BB	BB	BB	BB	BB	BB-BB	BB	BB	BB	BB	BB	BB	BB
1105	BB	BB	BB	BB	BB	BB	BB	BB-BB	BB	BB	BB	BB	BB	BB	BB
1121	BB	BB	BB	BB	BB	BB	BB	BB-BB	BB	BB	BB	00	00	BB	BB
1137	BB	BB	BB	BB	BB	BB	BB	BB-BB	BB	BB	BB	BB	BB	BB	BB
1153	BB	BB	BB	BB	BB	BB	BB	BB-BB	BB	BB	BB	BB	BB	BB	BB
1169	BB	BB	BB	BB	BB	BB	BB	BB-BB	BB	BB	BB	BB	BB	BB	BB
1185	BB	BB	BB	BB	00	00	BB	BB-BB	BB	BB	BB	BB	BB	BB	BB
1201	BB	BB	BB	BB	BB	BB	BB	BB-BB	BB	BB	BB	BB	BB	BB	BB
1217	BB	BB	BB	BB	BB	BB	BB	BB-BB	BB	BB	BB	BB	BB	BB	BB
1233	BB	BB	BB	BB	BB	BB	BB	BB-BB	BB	BB	BB	00	00	BB	BB
1249	BB	BB	BB	BB	BB	BB	BB	BB-BB	BB	BB	BB	BB	BB	BB	BB
1265	BB	BB	BB	BB	BB	BB	BB	BB-BB	BB	BB	BB	BB	BB	BB	BB
1281	BB	BB	BB	BB	BB	BB	BB	BB-BB	BB	BB	BB	BB	BB	BB	BB
1297	BB	BB	BB	BB	00	00	BB	BB-BB	BB	BB	BB	BB	BB	BB	BB
	Some	data	deleted	here	to	save	space								
5201	BB	BB	BB	BB	BB	BB	BB	BB-BB	BB	BB	BB	BB	BB	BB	BB
5217	BB	BB	BB	BB	00	00	BB	BB-BB	BB	BB	BB	BB	BB	BB	BB
5233	BB	BB	BB	BB	BB	BB	BB	BB-BB	BB	BB	BB	BB	BB	BB	BB
5249	BB	BB	BB	BB	BB	BB	BB	BB-BB	BB	BB	BB	BB	BB	BB	BB
5265	BB	BB	BB	BB	BB	BB	BB	BB-BB	BB	BB	BB	00	00	BB	BB
5281	BB	BB	BB	BB	BB	BB	BB	BB-BB	BB	BB	BB	BB	BB	BB	BB
5297	BB	BB	BB	BB	BB	BB	BB	BB-BB	BB	BB	BB	BB	BB	BB	BB
5313	BB	BB	BB	BB	BB	BB	BB	BB-BB	BB	BB	BB	BB	BB	BB	BB
5329	BB	BB	BB	BB	00	00									

Worksheet 11

Storage of data, bitmaps

The hex dump below is the top of a bitmapped file. Look at the values in the file header and fill in the correct values in Table 2.5.

Table 2.4 *Bitmap image file format details*

The bitmap file header

Start position	Size in bytes	Meaning or use
1	2	set to 'BM' to show this is a .bmp file.
3	4	size of the file in bytes.
7	2	always set to zero.
9	2	always set to zero.
11	4	offset from the beginning of the file to the bitmap data.

The bitmap information header

15	4	size of the bitmap information header structure, in bytes.
19	4	width of the image, in pixels.
23	4	height of the image, in pixels.
27	2	set to zero.
29	2	number of bits per pixel.
31	4	compression, usually set to zero, 0 = no compression.
35	4	size of the image data, in bytes.
39	4	horizontal pixels per metre, usually set to zero.
43	4	vertical pixels per metre, usually set to zero.
47	4	number of colors used in the bitmap.
51	4	number of colours that are 'important' for the bitmap if set to zero, all colours are important.

```
42  4D  36  1B  B7  00  00  00  00  00  36  00  00  00  28  00    BM6.•.....6...(.
00  00  D0  07  00  00  D0  07  00  00  01  00  18  00  00  00    ..Ð...Ð........
00  00  00  1B  B7  00  66  5C  00  00  66  5C  00  00  00  00    .....•.f\..f\......
00  00  00  00  00  00  FF  FF  FF  FF  FF  FF  FF  FF  FF  FF    ......ÿÿÿÿÿÿÿÿÿÿ
FF  FF  FF  FF  FF  FF  FF  FF  FF  FF  FF  FF  FF  FF  FF  FF    ÿÿÿÿÿÿÿÿÿÿÿÿÿÿÿÿ
FF  FF  FF  FF  FF  FF  FF  FF  FF  FF  FF  FF  FF  FF  FF  FF    ÿÿÿÿÿÿÿÿÿÿÿÿÿÿÿÿ
FF  FF  FF  FF  FF  FF  FF  FF  FF  FF  FF  FF  FF  FF  FF  FF    ÿÿÿÿÿÿÿÿÿÿÿÿÿÿÿÿ
FF  FF  FF  FF  FF  FF  FF  FF  FF  FF  FF  FF  FF  FF  FF  FF    ÿÿÿÿÿÿÿÿÿÿÿÿÿÿÿÿ
FF  FF  FF  FF  FF  FF  FF  FF  FF  FF  FF  FF  FF  FF  FF  FF    ÿÿÿÿÿÿÿÿÿÿÿÿÿÿÿÿ
FF  FF  FF  FF  FF  FF  FF  FF  FF  FF  FF  FF  FF  FF  FF  FF    ÿÿÿÿÿÿÿÿÿÿÿÿÿÿÿÿ
FF  FF  FF  FF  FF  FF  FF  FF  FF  FF  FF  FF  FF  FF  FF  FF    ÿÿÿÿÿÿÿÿÿÿÿÿÿÿÿÿ
FF  FF  FF  FF  FF  FF  FF  FF  FF  FF  FF  FF  FF  FF  FF  FF    ÿÿÿÿÿÿÿÿÿÿÿÿÿÿÿÿ
```

Table 2.5 *In decimal, what is the*

	Hex values, little endian	Hex values, big endian	Decimal value
Image file size			
Image width			
Image height			
Number of bits per pixel			
Max. number of colours possible	n/a	n/a	

Worksheet 12

Storage of data, bitmaps

The hex dump below is the top of a bitmapped file. Look at the values in the file header and fill in the correct values in Table 2.7.

Table 2.6 *Bitmap image file format details*

The bitmap file header

Start position	Size in bytes	Meaning or use
1	2	set to 'BM' to show this is a .bmp file.
3	4	size of the file in bytes.
7	2	always set to zero.
9	2	always set to zero.
11	4	offset from the beginning of the file to the bitmap data.

The bitmap information header

15	4	size of the bitmap information header structure, in bytes.
19	4	width of the image, in pixels.
23	4	height of the image, in pixels.
27	2	set to zero.
29	2	number of bits per pixel.
31	4	compression, usually set to zero, 0 = no compression.
35	4	size of the image data, in bytes.
39	4	horizontal pixels per metre, usually set to zero.
43	4	vertical pixels per metre, usually set to zero.
47	4	number of colours used in the bitmap.
51	4	number of colours that are 'important' for the bitmap; if set to zero, all colours are important.

```
42  4D  8E  37  38  00  00  00  00  00  36  04  00  00  28  00   BM□78.....6...(.
00  00  22  09  00  00  26  06  00  00  01  00  08  00  00  00   .."...&.........
00  00  58  33  38  00  F2  1E  00  00  F2  1E  00  00  00  01   ..X38.ò...ò.....
00  00  00  01  00  00  45  47  59  00  7C  4D  55  00  50  76   ......EGY.|MU.Pv
66  00  84  75  5B  00  37  65  86  00  6C  64  84  00  82  59   f.„u[.7e†.ld„.,Y
87  00  84  7C  81  00  27  67  A5  00  29  8A  A3  00  53  5B   ‡.„|□.'g¥.)Š£.S[
A7  00  53  87  A4  00  73  54  A7  00  74  6C  AA  00  77  80   §.S‡ª.sT§.tlª.w□
94  00  75  81  B0  00  A8  59  45  00  B3  79  4B  00  A7  5D   ".u□°."YE.³yK.§]
6E  00  B0  7A  6B  00  9A  67  7D  00  B8  6B  7E  00  A3  83   n.°zk.šg}.,k~.£ƒ
78  00  C2  7F  78  00  8E  68  8E  00  8C  6C  AB  00  AB  6E   x.Â▓x.□h□.Œl«.«n
8D  00  A5  73  A6  00  98  88  8D  00  95  88  AB  00  BD  84   □.¥s¦.˜^.•^«.½„
84  00  B9  84  A4  00  27  97  A1  00  32  B5  A4  00  5A  A7   „.¹„ª.'—¡.2µ€.Z§
```

Table 2.7 *In decimal, what is the*

	Hex values, little endian	Hex values, big endian	Decimal value
Image file size			
Image width			
Image height			
Number of bits per pixel			
Max. number of colours possible	n/a	n/a	

Worksheet 13

Storage of data, bitmaps

The hex dump below is the top of a bitmapped file. Look at the values in the file header and fill in the correct values in Table 2.9.

Table 2.8 *Bitmap image file format details*

The bitmap file header

Start position	Size in bytes	Meaning or use
1	2	set to 'BM' to show this is a .bmp file.
3	4	size of the file in bytes.
7	2	always set to zero.
9	2	always set to zero.
11	4	offset from the beginning of the file to the bitmap data.

The bitmap information header

15	4	size of the Bitmap information header structure, in bytes.
19	4	width of the image, in pixels.
23	4	height of the image, in pixels.
27	2	set to zero.
29	2	number of bits per pixel.
31	4	compression, usually set to zero, 0 = no compression.
35	4	size of the image data, in bytes.
39	4	horizontal pixels per metre, usually set to zero.
43	4	vertical pixels per metre, usually set to zero.
47	4	number of colours used in the bitmap.
51	4	number of colours that are 'important' for the bitmap; if set to zero, all colours are important.

```
42 4D 36 96 10 00 00 00 00 00 76 00 00 00 28 00   BM6-......v...(.
00 00 C4 04 00 00 F0 06 00 00 01 00 04 00 00 00   ..Ä...ð.........
00 00 C0 95 10 00 43 2E 00 00 43 2E 00 00 10 00   ..À•..C...C.....
00 00 10 00 00 00 47 50 50 00 5A 68 63 00 5B 6B   ......GPP.Zhc.[k
70 00 67 6E 6F 00 62 73 77 00 63 78 82 00 72 7B   p.gno.bsw.cx,.r{
7E 00 76 87 8D 00 81 94 9D 00 8A A3 AD 00 9C AC   ~.v‡□.□"□.Š£-.œ¬
B3 00 A4 B8 C0 00 9F BD CA 00 AC BF CC 00 B2 CD   ³.¤¸À.Ÿ½Ê.¬¿Ì.²Í
D4 00 C1 DE EC 00 66 66 46 44 66 77 64 66 66 66   Ô.ÁÞì.ffFDfwdfff
66 66 66 66 66 66 54 66 66 66 66 66 66 44 44 43   ffffffTffffffDDC
33 22 33 33 33 33 34 44 43 33 33 22 23 33 33 33   3"33334DC33"#333
33 33 34 44 66 44 44 44 44 44 44 44 44 66 44 44   334DfDDDDDDDfDD
44 44 46 66 66 66 66 66 66 66 44 44 44 44 44 44   DDFfffffffDDDDDD
```

Table 2.9 *In decimal, what is the*

	Hex values, little endian	Hex values, big endian	Decimal value
Image file size			
Image width			
Image height			
Number of bits per pixel			
Max. number of colours possible	n/a	n/a	

Worksheet 14

Storage of data, bitmaps

The hex dump below is the top of a bitmapped file. Look at the values in the file header and fill in the correct values in Table 2.11.

Table 2.10 *Bitmap image file format details*

The bitmap file header

Start position	Size in bytes	Meaning or use
1	2	Set to 'BM' to show this is a .bmp file.
3	4	size of the file in bytes.
7	2	always set to zero.
9	2	always set to zero.
11	4	offset from the beginning of the file to the bitmap data.

The bitmap information header

15	4	size of the bitmap information header structure, in bytes.
19	4	width of the image, in pixels.
23	4	height of the image, in pixels.
27	2	set to zero.
29	2	number of bits per pixel.
31	4	compression, usually set to zero, 0 = no compression.
35	4	size of the image data, in bytes.
39	4	horizontal pixels per metre, usually set to zero.
43	4	vertical pixels per metre, usually set to zero.
47	4	number of colours used in the bitmap.
51	4	number of colours that are 'important' for the bitmap; if set to zero, all colours are important.

```
42 4D 62 D6 01 00 00 00 00 00 36 04 00 00 28 00   BMbÖ......6...(.
00 00 52 01 00 00 5F 01 00 00 01 00 08 00 00 00   ..R..._.........
00 00 2C D2 01 00 23 2E 00 00 23 2E 00 00 00 01   ..,Ò..#...#.....
00 00 00 01 00 00 00 00 00 00 01 01 01 00 02 02   ................
02 00 03 03 03 00 04 04 04 00 05 05 05 00 06 06   ................
06 00 07 07 07 00 08 08 08 00 09 09 09 00 0A 0A   ................
0A 00 0B 0B 0B 00 0C 0C 0C 00 0D 0D 0D 00 0E 0E   ................
0E 00 0F 0F 0F 00 10 10 10 00 11 11 11 00 12 12   ................
12 00 13 13 13 00 14 14 14 00 15 15 15 00 16 16   ................
16 00 17 17 17 00 18 18 18 00 19 19 19 00 1A 1A   ................
1A 00 1B 1B 1B 00 1C 1C 1C 00 1D 1D 1D 00 1E 1E   ................
1E 00 1F 1F 1F 00 20 20 20 00 21 21 21 00 22 22   ......   .!!!."" 
```

Table 2.11 *In decimal, what is the*

	Hex values, little endian	Hex values, big endian	Decimal value
Image file size			
Image width			
Image height			
Number of bits per pixel			
Max. number of colours possible	n/a	n/a	

Worksheet 15

Storage of data, bitmaps

The hex dump below is the top of a bitmapped file. Look at the values in the file header and fill in the correct values in Table 2.13.

Table 2.12 *Bitmap image file format details*

The bitmap file header

Start position	*Size in bytes*	*Meaning or use*
1	2	Set to 'BM' to show this is a .bmp file.
3	4	size of the file in bytes.
7	2	always set to zero.
9	2	always set to zero.
11	4	offset from the beginning of the file to the bitmap data.

The bitmap information header

15	4	size of the bitmap information header structure, in bytes.
19	4	width of the image, in pixels.
23	4	height of the image, in pixels.
27	2	set to zero.
29	2	number of bits per pixel.
31	4	compression, usually set to zero, 0 = no compression.
35	4	size of the image data, in bytes.
39	4	horizontal pixels per metre, usually set to zero.
43	4	vertical pixels per metre, usually set to zero.
47	4	number of colours used in the bitmap.
51	4	number of colours that are 'important' for the bitmap; if set to zero, all colours are important.

```
42 4D 36 50 1D 00 00 00 00 00 36 04 00 00 28 00  BM6P......6...(.
00 00 40 06 00 00 B0 04 00 00 01 00 08 00 00 00  ..@...°.........
00 00 00 4C 1D 00 13 0B 00 00 13 0B 00 00 00 01  ...L............
00 00 00 01 00 00 00 00 00 00 01 01 01 00 02 02  ................
02 00 03 03 03 00 04 04 04 00 05 05 05 00 06 06  ................
06 00 07 07 07 00 08 08 08 00 09 09 09 00 0A 0A  ................
0A 00 0B 0B 0B 00 0C 0C 0C 00 0D 0D 0D 00 0E 0E  ................
0E 00 0F 0F 0F 00 10 10 10 00 11 11 11 00 12 12  ................
12 00 13 13 13 00 14 14 14 00 15 15 15 00 16 16  ................
16 00 17 17 17 00 18 18 18 00 19 19 19 00 1A 1A  ................
1A 00 1B 1B 1B 00 1C 1C 1C 00 1D 1D 1D 00 1E 1E  ................
1E 00 1F 1F 1F 00 20 20 20 00 21 21 21 00 22 22  ......   .!!!.""
```

Table 2.13 *In decimal, what is the*

	Hex values, little endian	*Hex values, big endian*	*Decimal value*
Image file size	36 50 1D 00	00 1D 50 36	1 921 078
Image width	40 06 00 00	00 00 06 40	1600
Image height	B0 04 00 00	00 00 04 B0	1200
Number of bits per pixel	08 00	00 08	8
Max. number of colours possible	n/a	n/a	256

Worksheet 16

Storage of data, bitmaps

The hex dump below is the top of a bitmapped file. Look at the values in the file header and fill in the correct values in Table 2.15.

Table 2.14 *Bitmap image file format details*

The bitmap file header

Start position	Size in bytes	Meaning or use
1	2	Set to 'BM' to show this is a .bmp file.
3	4	size of the file in bytes.
7	2	always set to zero.
9	2	always set to zero.
11	4	offset from the beginning of the file to the bitmap data.

The bitmap information header

15	4	size of the bitmap information header structure, in bytes.
19	4	width of the image, in pixels.
23	4	height of the image, in pixels.
27	2	set to zero.
29	2	number of bits per pixel.
31	4	compression, usually set to zero, 0 = no compression.
35	4	size of the image data, in bytes.
39	4	horizontal pixels per metre, usually set to zero.
43	4	vertical pixels per metre, usually set to zero.
47	4	number of colours used in the bitmap.
51	4	number of colours that are 'important' for the bitmap; if set to zero, all colours are important.

```
42  4D  C2  86  20  00  00  00  00  00  36  00  00  00  28  00   BMÂ† .....6...(.
00  00  CB  03  00  00  DB  02  00  00  01  00  18  00  00  00   ..Ë...Û.........
00  00  8C  86  20  00  30  5C  00  00  30  5C  00  00  00  00   ..Œ† .0\..0\....
00  00  00  00  00  00  31  45  4A  31  45  4A  31  45  4A  31   ......1EJ1EJ1EJ1
45  4A  31  45  4A  39  49  4A  39  4D  52  39  4D  52  39  4D   EJ1EJ9IJ9MR9MR9M
52  39  4D  52  39  4D  52  39  4D  52  39  4D  52  39  49  52   R9MR9MR9MR9MR9IR
39  49  52  39  49  4A  31  49  4A  31  49  52  31  49  52  31   9IR9IJ1IJ1IR1IR1
49  52  39  4D  52  39  55  5A  42  59  5A  42  59  5A  39  55   IR9MR9UZBYZBYZ9U
5A  42  59  5A  42  59  63  39  55  5A  39  55  5A  39  55  5A   ZBYZBYc9UZ9UZ9UZ
42  55  5A  42  51  5A  39  51  5A  42  55  5A  4A  5D  63  52   BUZBQZ9QZBUZJ]cR
69  73  5A  71  7B  63  79  84  63  79  84  63  79  84  5A  75   isZq{cy„cy„cy„Zu
7B  5A  75  7B  5A  75  7B  5A  71  7B  5A  71  7B  5A  6D  73   {Zu{Zu{Zq{Zq{Zms
```

Table 2.15 *In decimal, what is the*

	Hex values, little endian	Hex values, big endian	Decimal value
Image file size			
Image width			
Image height			
Number of bits per pixel			
Max. number of colours possible	n/a	n/a	

Worksheet 17

Number bases

Fill in the spaces in Table 2.16 by converting number bases as required.

Hints:

- Binary to hex or hex to binary, use groups of four binary digits.
- Binary to octal or octal to binary, use groups of three binary digits.
- If running Windows, you can use Start→Programs→Accessories→Calculator (with View set to Scientific) to check your answers. There is no point using the calculator to complete the conversions, you should do them on paper alone to get the most practice.

Table 2.16

Decimal	Hex	Octal	Binary
	1C9DC		
			010110010111
		2024	
4			
91 509			
		2340	
	45		
	18A0B		
	154		
66 182			
			0010001000001000
10 815			

Worksheet 18

Number bases

Fill in the spaces in Table 2.17 by converting number bases as required.

Hints:

- Binary to hex or hex to binary, use groups of four binary digits.
- Binary to octal or octal to binary, use groups of three binary digits.
- If running Windows, you can use Start→Programs→Accessories→Calculator (with View set to Scientific) to check your answers. There is no point using the calculator to complete the conversions, you should do them on paper alone to get the most practice.

Table 2.17

Decimal	Hex	Octal	Binary
			0001000100100000
4252			
		264767	
47			
			0010011111110101
		350	
589			
			0110100010110101
	DD		
			110001011000
	226D		
		675	

Worksheet 19

Number bases

Fill in the spaces in Table 2.18 by converting number bases as required.

Hints:

- Binary to hex or hex to binary, use groups of four binary digits.
- Binary to octal or octal to binary, use groups of three binary digits.
- If running Windows, you can use Start→Programs→Accessories→Calculator (with View set to Scientific) to check your answers. There is no point using the calculator to complete the conversions, you should do them on paper alone to get the most practice.

Table 2.18

Decimal	Hex	Octal	Binary
		24040	
			0001000111110111
		24104	
		20066	
377			
			101110000011
			0001100100101011
11 389			
9148			
	285C		
	1F0A3		
	D97		

Worksheet 20

Number bases

Fill in the spaces in Table 2.19 by converting number bases as required.

Hints:

- Binary to hex or hex to binary, use groups of four binary digits.
- Binary to octal or octal to binary, use groups of three binary digits.
- If running Windows, you can use Start→Programs→Accessories→Calculator (with View set to Scientific) to check your answers. There is no point using the calculator to complete the conversions, you should do them on paper alone to get the most practice.

Table 2.19

Decimal	Hex	Octal	Binary
	478E		
	1F0C		
	2086		
			0001100100101011
		2001	
93 208			
98 757			
		11176	
			1101101000100101
			0010110010111101
			0001110101010001
			11010010

Worksheet 21

Number bases

Fill in the spaces in Table 2.20 by converting number bases as required.

Hints:

- Binary to hex or hex to binary, use groups of four binary digits.
- Binary to octal or octal to binary, use groups of three binary digits.
- If running Windows, you can use Start→Programs→Accessories→Calculator (with View set to Scientific) to check your answers. There is no point using the calculator to complete the conversions, you should do them on paper alone to get the most practice.

Table 2.20

Decimal	Hex	Octal	Binary
	15E		
	297		
	1FE		
			011010101010
			0010001000100110
			0001000111001110
8702			
8458			
12 469			
		10260	
		115	
		12542	

Worksheet 22

Number bases

Fill in the spaces in Table 2.21 by converting number bases as required.

Hints:

- Binary to hex or hex to binary, use groups of four binary digits.
- Binary to octal or octal to binary, use groups of three binary digits
- If running Windows, you can use Start→Programs→Accessories→Calculator (with View set to Scientific) to check your answers. There is no point using the calculator to complete the conversions, you should do them on paper alone to get the most practice.

Table 2.21

Decimal	Hex	Octal	Binary
	3087		
	21EB		
	123B5		
112 061			
			000101100111
			11101001
2973			
123 366			
			00011101
		25032	
		1422	
		23667	

2.3 Logic gates

Worksheets 23 to 29 are presented to provide practice in understanding logic gates. Although Karnough maps and Boolean algebra have been left out, competence in problems of this complexity are sufficient for the National Certificate/Diploma in Computing up to E grade.

Worksheet 23

Logic gates

Determine the behaviour of the logic circuit shown in Figure 2.2. Use these steps:

- Fill in all the possible logic states of the inputs A, B and C.
- Label intermediate points in the circuit with D and E, i.e. the output of each gate.
- Label the output of the whole circuit as R.
- Fill in the truth table for D and E.
- Fill in the resultant R.

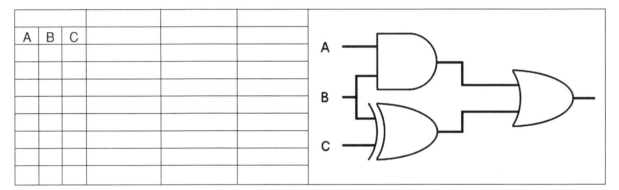

Figure 2.2 *Logic circuit and truth table*

Worksheet 24

Logic gates

Determine the behaviour of the logic circuit shown in Figure 2.3. Use these steps:

- Fill in all the possible logic states of the inputs A, B and C.
- Label intermediate points in the circuit with D and E, i.e. the output of each gate.
- Label the output of the whole circuit as R.
- Fill in the truth table for D and E.
- Fill in the resultant R.

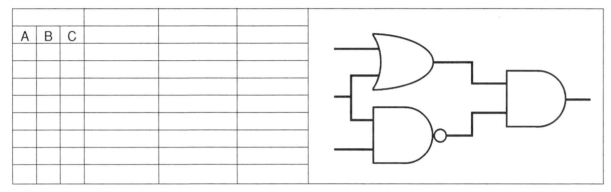

A	B	C				

Figure 2.3 *Logic circuit and truth table*

Worksheet 25

Logic gates

Determine the behaviour of the logic circuit shown in Figure 2.4. Use these steps:

- Fill in all the possible logic states of the inputs A, B and C.
- Label intermediate points in the circuit with D and E, i.e. the output of each gate.
- Label the output of the whole circuit as R.
- Fill in the truth table for D and E.
- Fill in the resultant R.

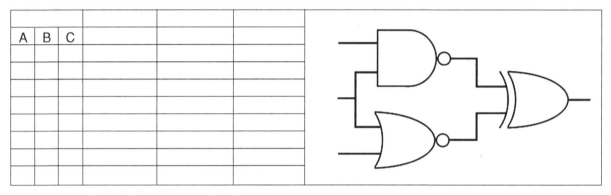

Figure 2.4 *Logic circuit and truth table*

Worksheet 26

Logic gates

Determine the behaviour of the logic circuit shown in Figure 2.5. Use these steps:

- Fill in all the possible logic states of the inputs A, B and C.
- Label intermediate points in the circuit with D and E, i.e. the output of each gate.
- Label the output of the whole circuit as R.
- Fill in the truth table for D and E.
- Fill in the resultant R.

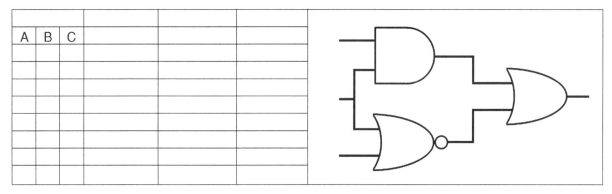

A	B	C			

Figure 2.5 *Logic circuit and truth table*

Worksheet 27

Logic gates

Determine the behaviour of the logic circuit shown in Figure 2.6. Use these steps:

- Fill in all the possible logic states of the inputs A, B, C and D.
- Label intermediate points in the circuit with E, F, G and H, i.e. the output of each gate.
- Label the output of the whole circuit as R.
- Fill in the truth table for E, F, G and H.
- Fill in the resultant R.

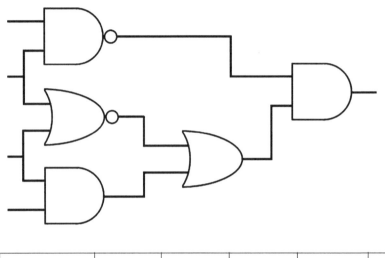

A	B	C	D					

Figure 2.6 *Logic circuit and truth table*

Worksheet 28

Logic gates

Determine the behaviour of the logic circuit shown in Figure 2.7. Use these steps:

- Fill in all the possible logic states of the inputs A, B and C.
- Label intermediate points in the circuit with D, E and F, i.e. the output of each gate.
- Label the output of the whole circuit as R.
- Fill in the truth table for D, E and F.
- Fill in the resultant R.

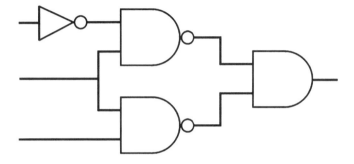

A	B	C					

Figure 2.7 *Logic circuit and truth table*

Worksheet 29

Logic gates

There is a mistake in the circuit diagram in Figure 2.8. Although the resultant R is correct in the truth table below, one of the truth table columns has the wrong heading and behaviour. One of the gates needs to be changed to a different kind. Which gate is incorrect?

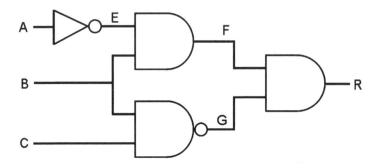

			NOT A	D NAND B	E AND B	B NAND C	F AND G
A	B	C	D	E	F	G	R
0	0	0	1	1	0	1	1
0	0	1	1	1	0	1	1
0	1	0	1	0	0	1	1
0	1	1	1	0	0	0	0
1	0	0	0	1	0	1	1
1	0	1	0	1	0	1	1
1	1	0	0	1	1	1	0
1	1	1	0	1	1	0	0

Figure 2.8 *Logic circuit and truth table*

2.4 Floating point numbers

Floating point numbers are very useful for the convenient representation of fractions but they are not always accurate.

Generally, decimal numbers whose fixed point binary presentation fits inside the number of bytes used by the floating point format will be accurate, i.e. there will be no loss of bits. Numbers such as 0.01 do not fit into this category as 0.01 decimal in fixed point binary is approximately 0.0000001010001110101110000101000111101011100001010000001110110 to 64 binary places or 0.000000101000111101011100001010001111010111000010 to 48 binary places.

If the floating point representation used, say, 48 bits, the last $64 - 48 = 16$ bits would be lost, i.e. the number is less accurate. Although this is true, it is important to get it in perspective. The difference, in decimal, between a 48-bit and 64-bit representation of decimal 0.01 is just 1.810^{-15}!

Applications where large numbers of iterations are performed are likely to show errors. An iteration is an operation that is done over and over again.

Consider this loop in Pascal, assuming the variable x is of type real:

```
x:=0;
repeat
        something(x);
        {call procedure called "something".
        What it is, is of no importance here}
        x:=x+0.001;
until x=1000;
```

This should execute 1 000 000 times but the loop may well not actually finish, it may go on forever (i.e. it may be infinite loop). This is because x may never equal 1000, it will simply be very close. The 'bug' is fixed by using the condition until x>=1000 this will terminate.

The better way to write this loop would be to use :=

```
for i:=0 to 1000000 do
        begin
                x:=i / 1000;
                something(x);
        end;
```

where i is of type *longint*. The loop will execute a known number of times and there is no cumulative error. Although the floating point calculation x:=i / 1000; will produce an error, this error will always be very small. The previous example added any error 1 000 000 times, so in effect 'amplifying' the error.

Longints can be from –2 147 483 648 to 2 147 483 647 because they are signed 32-bit numbers.

Worksheet 30

Floating point fractional numbers

You can convert a decimal fraction to a binary fraction to any number of places. The method is to successively multiply by 2 and to take the integer or whole part of the answer as a binary digit.

For example, convert 0.5703125 to a fixed point binary fraction.

0.5703125	× 2 =	1.140625	integer part =	1
0.140625	× 2 =	0.28125	integer part =	0
0.28125	× 2 =	0.5625	integer part =	0
0.5625	× 2 =	1.125	integer part =	1
0.125	× 2 =	0.25	integer part =	0
0.25	× 2 =	0.5	integer part =	0
0.5	× 2 =	1	integer part =	1

Answer = 0.1001001 reading from the top.

Use a spreadsheet to convert the following numbers:

- to at least 60 binary places
- if the numbers were then to be used in a floating point number with a 48-bit mantissa, show which of the numbers would cause a slight inaccuracy

Decimal	60 place binary	48-bit accurate?
0.1		
0.25678		
0.0625		
0.251220703		
0.141592654		

2.5 Assembly language

Assemble a program to mutiply a 16-bit number by an 8-bit number giving a 16 bit result.

Assuming the program below has been created as an ASCII file using a text editor, saved as trp1.asm and that the library file lib.asm is in the same directory, the command line to assemble trp1.asm to trp1.com is:

```
C:\> A86 trp1.asm lib.asm

mov ax,34      ;number to be multiplied
mov cl,2       ;number to multiply by
mul cl         ;multiply AX by contents if CL
call outint    ;show contents of AX with library routine
int 020        ;exit to DOS
```

You may find that the standard Windows text editor, Notepad, does not provide all the features that you may like. The best programmer's editor is called Ultraedit – details can be obtained from http://www.ultraedit.com/. This program will allow editing in hex and has a host of other features of use to programmers.

Hex dump of assembled .com file using DEBUG

The .com file is just 12 bytes long, shown in bold. The rest of the data given by DEBUG is just 'snow', i.e. irrelevant data.

```
0E95:0100   B8 22 00 B1 02 F6 E1 E8-42 00 CD 20 00 00 00 00   ."......B.. ....
0E95:0110   20 20 20 20 20 00 00 00-B4 02 8A 14 80 FA 00 74            .........t
0E95:0120   05 CD 21 46 EB F4 C3 8B-F0 B3 10 8B CE 81 E1 00   ..!F............
0E95:0130   80 83 F9 00 74 09 B2 31-B4 02 CD 21 E9 06 00 B2   ....t..1...!....
0E95:0140   30 B4 02 CD 21 D1 E6 FE-CB 75 E0 C3 B9 05 00 BE   0...!....u......
0E95:0150   10 01 B2 20 88 14 46 E2-FB BE 14 01 BA 00 00 B9   ... ..F........
0E95:0160   0A 00 F7 F1 A3 0E 01 80-CA 30 88 14 4E A1 0E 01   .........0..N...
0E95:0170   3D 00 00 75 E7 BE 10 01-E8 01 00 C3 B4 02 8A 14   =..u............
```

Unassembly using DEBUG

```
0E95:0100   B82200    MOV    AX,0022
0E95:0103   B102      MOV    CL,02
0E95:0105   F6E1      MUL    CL
0E95:0107   E84200    CALL   014C
0E95:010A   CD20      INT    20
```

Points to note

- The MUL instruction works with the AX register, the answer is placed back in the AX register.
- The symbolic address OUTINT has been converted to an absolute address 014C by the assembler.
- The program text mov ax,34 appears as MOV AX, 0022. The case is of no significance but DEBUG outputs in hex, 34 decimal is 22 hex. The .com file contains the bytes B8 22 00 which translate to MOV AX, 0022, the system stores numbers in *little endian* format, so 00 22 is the hex value 0022 or just 22.
- The OUTINT subroutine in LIB.ASM outputs in decimal and the program correctly ouputs the value 68.

Table 2.22 *Simplified subset of 8086 instruction set*

Op-code	Example	Comment
DD	ADD AX,12	Add 12 to what is in AX
AND	AND DH, 10011011b	Logical AND with DH and binary value 10011011
CALL	CALL 2000	Call routine at address 2000. It will end with a RET instruction
CMP	CMP SI,12	Compare SI register with value 12. Result sets flag in flag register
DEC	DEC AL	Subtract 1 from AL register
DIV	DIV DL	Divide AX by what is in DL. Answer, AL = Quotient, AH = Remainder
INC	INC SI	Add 1 to what is in SI register
INT	INT 021	Call DOS interrupt 21 (21 hex)
JA	JA 2000	Jump if above to address 2000
JAE	JAE 2100	Jump if above or equal to address 2100
JB	JB 3000	Jump if below to address 3000
JBE	JBE 1000	Jump if below or equal to address 1000
JC	JC 1000	Jump if carry flag set to address 1000
JMP	JMP 2300	Jump unconditionally to address 2300
JNZ	JNZ 2000	Jump if not zero to address 2000
MOV	MOV AX,23	Move 23 into AX register
MUL	MUL CL	Multiply AX register by what is in CL, answer in AX
OR	OR DI, 00110001b	Logical OR DI register with binary value 00110001
POP	POP AX	Take value from stack, put into AX, decrement stack pointer
PUSH	PUSH AX	Take value from AX, place on stack, increment stack pointer
RET	RET	Return from subroutine
SHL	SHL AX,3	Shift the value in AX 3 times to the left
SHR	SHR AX,3	Shift the value in AX 3 times to the right
SUB	SUB AX,8	Subtract 8 from value in AX register
XOR	XOR AL, AL	Logical eXclusive OR of AL with itself (sets AL to zero)

Assembler tasks

The following tasks can all be carried out using A86 and 8086 code, without the need to use extended registers etc. The aim is to provide a working knowledge of the *ideas* of assembler rather than to achieve competence in writing useful assembly code for modern machines. Many centres will have extensive security code to protect the settings of their operating systems and networks; these settings often preclude the use of assemblers and tools aimed at low level work such as DEBUG. For this reason, centres may choose to run older non-networked PCs for assembly practice and low level work. A86, DEBUG and the 8086 op-code set are ideal for these tasks on older machines.

Worksheet 31

Assembler

For the assembler tasks in worksheets 31 to 35, use can be made of the library file of assembly routines, LIB.ASM, available from http://www.bh.com/companions/0750656840 or copied from Appendix D. An ASCIIZ string is a simple string that uses ASCII character 0 as a terminator, i.e. the last byte of the string. This style of string is common in C programs etc.

```
;PRINTSTRING prints a string pointed to with the SI register
;OUTINT writes a 16 bit number in AX to the screen as integer
;PRINTNUM writes a string as PRINTSTRING but with no spaces
;SHOWBITS writes a 16 bit number in AX to the screen as binary
;leadingzero writes a single 0 to the screen
;crlf writes an ASCII 13 Carriage Return then ASCII 10 Line Feed
```

Write an assembler program that has these features:

- An ASCIIZ string embedded in the program with a maximum length of 254 bytes.
- The program outputs the string in lowercase.
- Presents the code with plenty of sensible comments.

Worksheet 32

Assembler

For the assembler tasks in worksheets 31 to 35, use can be made of the library file of assembly routines, LIB.ASM, available from http://wqw.bh.com/companions/0750656840 or copied from Appendix D. An ASCIIZ string is a simple string that uses ASCII character 0 as a terminator, i.e. the last byte of the string. This style of string is common in C programs etc.

```
;PRINTSTRING prints a string pointed to with the SI register
;OUTINT writes a 16 bit number in AX to the screen as integer
;PRINTNUM writes a string as PRINTSTRING but with no spaces
;SHOWBITS writes a 16 bit number in AX to the screen as binary
;leadingzero writes a single 0 to the screen
;crlf writes an ASCII 13 Carriage Return then ASCII 10 Line Feed
```

Write an assembler program that has these features:

- Reads two single numeric keys from the keyboard.
- Displays the '=' sign then the two keys added together.

Worksheet 33

Assembler

For the assembler tasks in worksheets 31 to 35, use can be made of the library file of assembly routines, LIB.ASM, available from http://www.bh.com/companions/0750656840 or copied from Appendix D. An ASCIIZ string is a simple string that uses ASCII character 0 as a terminator, i.e. the last byte of the string. This style of string is common in C programs etc.

```
;PRINTSTRING prints a string pointed to with the SI register
;OUTINT writes a 16 bit number in AX to the screen as integer
;PRINTNUM writes a string as PRINTSTRING but with no spaces
;SHOWBITS writes a 16 bit number in AX to the screen as binary
;leadingzero writes a single 0 to the screen
;crlf writes an ASCII 13 Carriage Return then ASCII 10 Line Feed
```

Write an assembler program that has these features:

- An ASCIIZ string embedded in the program.
- Returns a string with only characters in the range a–z, all other characters to be stripped out.

Worksheet 34

Assembler

For the assembler tasks in worksheets 31 to 35, use can be made of the library file of assembly routines, LIB.ASM, available from http://www.bh.com/companions/0750656840 or copied from Appendix D. An ASCIIZ string is a simple string that uses ASCII character 0 as a terminator, i.e. the last byte of the string. This style of string is common in C programs etc.

```
;PRINTSTRING prints a string pointed to with the SI register
;OUTINT writes a 16 bit number in AX to the screen as integer
;PRINTNUM writes a string as PRINTSTRING but with no spaces
;SHOWBITS writes a 16 bit number in AX to the screen as binary
;leadingzero writes a single 0 to the screen
;crlf writes an ASCII 13 Carriage Return then ASCII 10 Line Feed
```

Write an assembler program that reads a single key from the keyboard, then displays an '=' character, then its ASCII value. If you pressed the 'a' key, the program will display a=97.

The ASCII value will have two digits so use could be made of the library routine OUTINT in LIB.ASM.

You will need the DOS interupt as below.

DOS interupt 021 function 1, read keyboard with echo

Calling registers
      AH = 1
Return registers
      AL = ASCII value of key

Worksheet 35

Assembler

For the assembler tasks in worksheets 31 to 35, use can be made of the library file of assembly routines, LIB.ASM, available from http://www.bh.com/companions/0750656840 or copied from Appendix D. An ASCIIZ string is a simple string that uses ASCII character 0 as a terminator, i.e. the last byte of the string. This style of string is common in C programs etc.

```
;PRINTSTRING prints a string pointed to with the SI register
;OUTINT writes a 16 bit number in AX to the screen as integer
;PRINTNUM writes a string as PRINTSTRING but with no spaces
;SHOWBITS writes a 16 bit number in AX to the screen as binary
;leadingzero writes a single 0 to the screen
;crlf writes an ASCII 13 Carriage Return then ASCII 10 Line Feed
```

Write an assembler program that displays the status of the Num lock key.

BIOS interrupt 016 function 012, Return Keyboard status

Calling registers
    AH = 012

Return registers
    AH = status flags
    AL = status flags

AL bit 0 = Right shift key pressed
AL bit 1 = Left shift key pressed
AL bit 2 = Either CTRL key pressed
AL bit 3 = Either ALT key pressed
AL bit 4 = Scroll lock on
AL bit 5 = Num lock on
AL bit 6 = Caps lock on
AL bit 7 = SysReq (System Request) key on (does not work on all DOS versions)

Unit 3   Business information systems

3.1   Data and information

Worksheet 1

Processing data

Data is a set of unorganized, random facts, which have little or no meaning until they have undergone a processing activity. The function of processing is to convert the raw data into meaningful information.

Figure 3.1 *Conversion of data into information*

There are various types of processing activity which take place with certain data sets under certain conditions.

1. Complete the table below by describing the processing activity and give an example of when each processing type could take place.

Processing activity	Description	Example of use
Sorting		
Selecting		
Calculating		
Merging		
Manipulating		

Worksheet 2

Identifying data and information

1. From the options below select which items are data and which are information and put them into two columns.

date of birth   **invoice**   **today's date**

invitation to a party   **application form**   **01110010**

sales forecast sheet   **balance sheet**

measurements for a carpet   **car registration number**

hospital number   **curriculum vitae**   **telephone directory**

2. Make a third column of any items which you are unsure of.
3. As a group discuss the items in the third column and state why you were unsure as to whether they were data items or information.

Worksheet 3

Tools for communication

Information can be categorized into a number of groups, some of which include:

*   Verbal
*   Physical
*   Visual

Verbal communication implies the spoken word, for example: *giving directions*. Verbal communication is probably the most frequent and common form of communication because people exchange thoughts, views and ideas every second of every day.

Physical communication is expressed through body language, sign language, gestures and any other bodily expression, for example: *greeting somebody with a handshake*.

Visual communication is of the eye, what you can see or read. With the growing popularity of the internet, media and advertising visual communication has become more flamboyant and in cases interactive. Examples of visual communication include *posters*, *e-mails*, *billboards* and *television advertising* etc.

For each category of communication a number of tools can be used to convey the information (Figure 3.2).

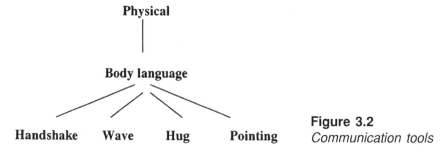

Figure 3.2
Communication tools

1. For each of the following scenarios identify a communication tool under the headings of verbal, physical or visual which would be most appropriate for the situation.
2. What alternative forms of communication could you use in the scenarios?
3. Are there any communication formats that you would not use in each scenario? If yes, why?

Scenario 1

You attended an interview three weeks ago and at the time you were told that they would let you know whether or not you had been successful within two weeks. Three weeks have passed and you have still not heard anything, how would you clarify the situation?

Scenario 2

In the heat of an argument with your best friend you said some things which you now regret. It has been three days since you last spoke, how could you now rectify the situation?

Scenario 3

You are the managing director of a small computing company. Over the past six months profits have been falling due to increased competition. It has now reached the point where you will have to make some employees redundant. The three people that will have to go have been with the company since it started eight years ago, how would you tell them?

3.2 Information in organizations

Worksheet 4

Identifying specialist information

Vast quantities of information are used and exchanged in organizations every second of every day. Information that is generated and communicated can be categorized under two headings:

- Generic
- Specialist

Generic information is based on general functional department and personnel needs and requirements, for example sales, marketing or finance. Specialist information develops from this but is unique to the organization. For example, organizations will develop their own marketing, pricing or IT strategies in-house.

1. For each of the three information diagrams input suitable specialist information requirements:

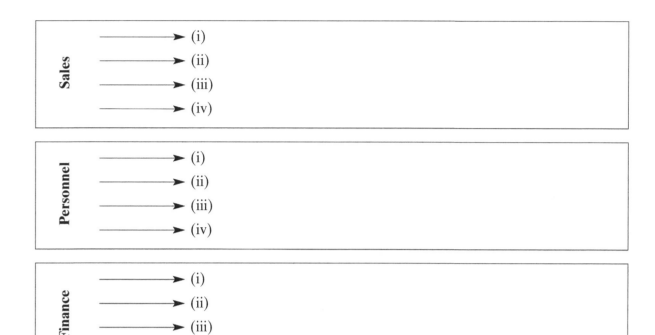

2. For the following organizations identify how the specialist information will change in the functional areas of operations, marketing and customer service.
    (i)   Charity
    (ii) Large insurance company
    (iii) Supermarket

3.3   Qualitative and quantitative information

Worksheet 5
Benefits and limitations of qualitative and quantitative information

Information can be described as being qualitative which implies that it is detailed, providing descriptions of a situation or event. Quantitative information is based upon figures and statistics, making the results easy to model and measure.

Look at each of the following three questionnaires – Film Poll, Product Ownership Form and Customer Services Satisfaction Sheet – and complete the following tasks:

1. State which one of the questionnaires is:
    (i)   Qualitative format
    (ii) Quantitative format
    (iii) Both
2. Which of the three questionnaires do you prefer in terms of structure and questioning?
3. Identify which questionnaire/s could be used to prepare statistical information using charts and graphs?
4. Identify two good points and two bad points for each questionnaire.
5. List four other questions which could be used for each questionnaire, keeping in with the adopted qualitative or quantitative format.

Questionnaire one – Film Poll

<table>
<tr><td colspan="2" align="center">**Film Poll**</td></tr>
<tr>
<td>Name:

Address:</td>
<td>Age (please tick the appropriate box):

18 or below ☐   19–35 ☐   36–50 ☐

51–64 ☐   65+   ☐</td>
</tr>
<tr><td colspan="2">What types of film do you watch?</td></tr>
<tr><td colspan="2">Why do you watch them?</td></tr>
<tr><td colspan="2">How often do you rent a video or DVD? (please tick the appropriate box)

2–3 times a week ☐   once a week ☐   fortnightly ☐   monthly ☐   occasionally ☐</td></tr>
<tr><td colspan="2">How often do you go to the cinema? (please tick the appropriate box)

2–3 times a week ☐   once a week ☐   fortnightly ☐   monthly ☐   occasionally ☐</td></tr>
<tr><td colspan="2">What influences your decision to watch a film? (please rate from 1 to 5)
1 = lowest 5 = highest

Actors/Actresses starring in it   ☐
Subject matter   ☐
Reviews   ☐
Knowledge of the story-line   ☐
Selected by friends to see   ☐</td></tr>
</table>

Questionnaire two – Product Ownership Form

<table>
<tr>
<td colspan="2" align="center">Product Ownership Form</td>
</tr>
<tr>
<td>Name:</td>
<td>Age (please tick the appropriate box):</td>
</tr>
<tr>
<td rowspan="2">Address:</td>
<td>18 or below ☐   19–35 ☐   36–50 ☐</td>
</tr>
<tr>
<td>51–64 ☐   65+   ☐</td>
</tr>
<tr>
<td>Do you work? ☐ Yes ☐ No
(If Yes)

Full-time ☐

Part-time ☐

Self-employed ☐</td>
<td>What is your salary per year?

Under £10 000   ☐   £10 000–£15 000 ☐

£15 000–£20 000 ☐   £20 000–£25 000 ☐

£25 000–30 000 ☐   £30 000+   ☐</td>
</tr>
</table>

Do you own any of the following products?
Please tick

	Yes	No
• Video	☐	☐
• DVD player	☐	☐
• Games console	☐	☐
• Hi-fi equipment	☐	☐
• Camera	☐	☐
• Digital camera	☐	☐
• PC	☐	☐
• Digital television	☐	☐
• Widescreen television	☐	☐
• Camcorder	☐	☐

How much do you spend on audio/electrical equipment each year?

Up to £500 ☐   £500–£1000 ☐   £1000–£3000 ☐   £3000–£6000 ☐   £6000+ ☐

Where do you buy your audio/electrical equipment?

Specialist shop ☐   Electrical retailer ☐   Supermarket ☐   Internet ☐   Mail order ☐

Do you buy extended warranties or guarantees?

Always ☐ On more expensive items ☐ On less expensive items ☐ Occasionally ☐ Never ☐

Questionnaire three – Customer Services Satisfaction Sheet

Customer Services Satisfaction Sheet	
Name:	Address:
Customer number:	
Please answer each question in as much detail as possible:	

1.	In what ways have you been happy with the service you received from the after-sales team?
2.	In what ways have you been unhappy with the service you received from the after-sales team?
3.	Describe the general attitude of customer service staff members to you.
4.	How, if at all, could this service be improved?
5.	Explain what happened when you returned your faulty product to customer services.
6.	What actions were taken as a result of your complaint regarding the faulty product?

Would you consider buying other products from us in the future, please state your reasons?

Additional comments:

Worksheet 6

Designing questionnaires

1. Design a questionnaire to suit one of the following conditions:
    (i) What people watch on television
    (ii) Where people like to go on holiday
    (iii) Type of career wanted at the end of the course
    For each questionnaire include a mixture of ten qualitative and quantitative questions. When the design is complete print off or e-mail five copies to people within the group and get them to complete them.
2. From the information gathered produce one chart and one graph based on the quantitative information and provide a short summary detailing the findings of the qualitative information.
3. Identify which was the easier of the two sets of information to produce graphical analysis and which was easier to produce written analysis. Why do you think this is?

3.4 Systems

Systems can be very simple or quite complex depending upon the function of the system, resources to support the system and factors such as costs, system objective and the systems environment. Systems do not work in isolation, they are dependent on items feeding into it, activities taking place within it and items being output from it.

A typical systems model is based upon a number of components, these include:

- Inputs
- Processes
- Outputs
- Feedback
- Control

The majority of all systems will have a combination of these components regardless of its simplicity or complexity.

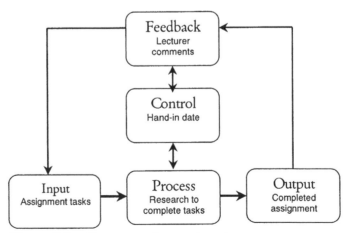

Figure 3.3 *Course assignment system*

For example, a system for doing your course assignments:

Inputs   Lecture notes, assignment brief
Process   Carry out research, build, design, complete task assignments
Outputs   Complete assignment
Feedback Written and/or verbal comments from lecturer
Control   Hand-in date for assignment

Worksheet 7

Examining systems

1. For the following systems identify the input/s, processes and output/s:
   (i) Making a cup of tea
   (ii) Buying a pair of trainers
   (iii) Getting up in the morning
   (iv) Learning to drive a car or ride a bike
2. For two of these systems identify what feedback and control mechanisms exist.
3. Identify three other systems and list the inputs, processes, outputs, feedback and control mechanisms which feed into making the systems complete.

Worksheet 8

System objectives

Systems are driven by an objective or objectives which can usually be measurable, for example the amount of profit made, distance travelled, achievements generated, mission accomplished.

1. You yourself are a system. Identify six system objectives which keep you functioning on a day-to-day basis and give you the willpower to study at college.
2. For each of the following systems identify the main objective/s:
   (i) Broadcasting the news daily
   (ii) Having a rota to do household chores
   (iii) Taking a driving test
   (iv) Having end of term exams
   (v) Attending an interview

System objectives may change depending upon the users involved with the system, so two or more people may have very different expectations of the system.

3. For systems 2(iii), (iv) and (v) identify system objectives from at least two users' points of view.

3.5   Organizations

Worksheet 9

Categories of information

Organizations can be categorized as being:

- Public sector
- Private sector
- Mutual
- Charity

1. Provide a description for each of the four categories of organization.
2. Complete the table by identifying an actual organization that would fit into each category and comparing it in terms of its function, what it provides, e.g. a product or a service, and how it is supported financially.

Organization type	Function	Product/Service	Financed
Public sector, e.g.			
Private sector, e.g.			
Mutual, e.g.			
Charity, e.g.			

Every organization has a structure, however informal, from corner shops to multinational corporations. The overall structure will depend upon a number of factors including:

- Size
- Amount of human resources
- Nature of business
- Scale of markets – local, national, international
- Collaborations

Worksheet 10

Organizational structures

Examine the two structures in Figures 3.4 and 3.5 and carry out the following tasks:

1. Label each structure.
2. Attach appropriate labels for each level of the organization structure.
3. Identify four organizations that would fit each structure.
4. Describe the benefits of each structure.

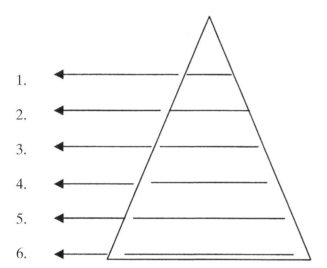

1.
2.
3.
4.
5.
6.

Figure 3.4 *Structure 1*

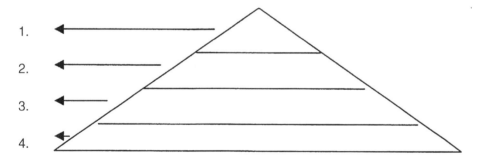

1.
2.
3.
4.

Figure 3.5 *Structure 2*

Worksheet 11

Types of decisions

As well as each organization having its own discrete employee and management levels, all organizations also work within a standard framework which is broken down into three overall levels. These levels can be linked to the types of decisions that organizations have to make in the short, medium and long term. The levels at which these decisions are made are:

- Strategic
- Tactical
- Operational

1. Describe the types of activities that take place within each of the three levels.
2. Identify the type of employee that would function within these three levels.
3. In terms of decision making, what period of planning would take place within each level?

Decisions that are made within an organization are not only confined to different levels but conform to various strata running through each level. These decisions can be identified in accordance with the nature of the decision to be made. For example, is the decision to do with routine processing and daily tasks, or is the decision more high level focusing on issues of acquiring new premises or targeting new markets? These three types of decision strata can be referred to as:

- Structured
- Semi-structured
- Unstructured

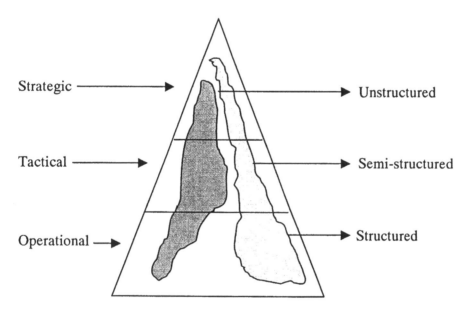

Figure 3.6 *Decision levels and decision strata*

Worksheet 12

Decision making

Complete the table to provide two examples of the types of decisions that could be made at unstructured, semi-structured and structured levels.

Type of decision	Example
Structured	1 2
Semi-structured	1 2
Unstructured	1 2

3.6   Functional areas

Worksheet 13

Functional departments

Organizations are built upon a collection of functions, the infrastructure of any organization being dependent upon the support provided by functional departments. Functional departments serve a specific purpose within the organization and together the departments enable the organization to achieve its objectives.

1.   Describe the key activities which take place in the following functional departments:
   (i)   Sales
   (ii)   Human Resources
   (iii) Finance
   (iv)   IT
   (v)   Marketing
   (vi)   Operations
   (vii) Customer services

2. State the reasons why the majority of organizations are organized by function. What are the benefits of this?

3. Identify which of the following activities and services belong to a particular functional department:

Department/s	Activity
	1. Recruiting a new network manager to the IT department
	2. Dealing with a customer complaint about an item that arrived damaged from sales
	3. Enforcing the 4% pay rise into employees' wage slips
	4. Setting up a new advertising campaign for the launch of a new product
	5. Ensuring that the new raw materials have been put through quality control
	6. Ringing a customer to inform them that their items are being despatched
	7. Ensuring software compatibility with the new network installation

4. What other functional departments exist within organizations? List and describe their purpose.

3.7 Information in organizations

Worksheet 14

Factors affecting information transmission

Information exchange in organizations occurs every second of every day, using a variety of formats, between different organizational levels and across many geographical boundaries.

The way in which information is transmitted both physically, using a specific transmission tool, and literally, in terms of how it is received and interpreted, depends upon a number of factors, these include:

*   The relationship between the sender and the receiver
*   The purpose or objective of the information exchange
*   The locality over which the information needs to be exchanged
*   The information priority, e.g. is it confidential or urgent?

- Its impact upon other information exchanges
- Technology

1. Using the first three factors provide an example within an organizational context of how each can affect the transmission of information.
2. How and why would technology impact upon the transmission of information? Give three examples of this.
3 If information had to be transmitted urgently within the hour and confidentially between two branches of an organization, which transmission tool/s would you use and why?

Information can be transmitted electronically, verbally or in a written/visual format within an organization.

Worksheet 15

Ways of transmitting information

1. Complete the table to identify at least three examples of information formats, identifying the benefits and limitations of each.

Information format	Benefits	Limitations
Electronic (i) (ii) (iii)		
Verbal (i) (ii) (iii)		
Written/Visual (i) (ii) (iii)		

3.8   Case study – The Finance Group Banking Corporation

Jack Humphries works for a large banking group. Within his particular branch there are six levels of employee. Margaret Fisher is the branch manager, she oversees the overall running of the bank and reports back to head office on a daily basis. Pete Moore is the assistant branch manager who works alongside Margaret and deputizes in her absence. Pete looks after the service managers of each department within the branch. There are six service managers specializing in the following:

- Pensions (1 team leader and 6 staff)
- Investments (1 team leader and 8 staff)
- Savings (1 team leader and 7 staff)
- Mortgages (1 team leader and 10 staff)
- Loans (1 team leader and 6 staff)
- New accounts (1 team leader and 9 staff)

Each service manager has an assistant manager who reports directly to them (with the exception of mortgages where there are two assistant managers because the department is continually growing, one for new customers and one for existing customers). The assistant managers are involved with operational issues such as enforcing new promotions, assisting the managers with short-term planning, providing facts and figures which are discussed at weekly management meetings and overseeing the day-to-day running of the team leaders and their teams. Each department has a team leader who oversees between six and ten people who have various roles such as bank clerk, customer advisor, cashier, or administrator.

Jack Humphries is the team leader for mortgages, he supervises ten people which can be a bit of a task, especially when he also has two bosses to report to, who are both quite demanding.

3.9   Flow of information in organizations

Worksheet 16

Communication flows

Information flows in a number of directions within an organization. This flow can include:

- Upwards
- Downwards
- Lateral

Upward flow implies that information is sourced at the lower organizational level and extends upwards. Downwards implies that information trickles from the upper organizational levels through to the lower. Lateral means that information is passed across on the same organizational levels.

Using the 'Finance Group Banking Corporation' case study in section 3.8, identify the flow of information from each person and draw an organizational structure to illustrate how the information will be passed through.

1. Draw an organizational chart to represent the structure of the bank.
2. What types of communication are the following situations?
    (i) Jack wants to communicate with the team leader in 'pensions'
    (ii) Margaret needs to feed information about weekly performance figures through to head office
    (iii) Pete needs to inform his service managers about the new change in interest rates
    (iv) Jack has to ring up another team leader at a different branch to discuss a new mortgage scheme
3. When Jack is on holiday nobody deputizes for him. Identify what problems might occur.
4. Jack has to report to two bosses, the assistant manager for new mortgage customers and the assistant manager for existing mortgage customers. Identify what problems this might cause.

3.10 Uses of information in organizations

Worksheet 17

Uses of information in organizations

Information is used for a variety of purposes in organizations including:

• Planning
• Forecasting
• Decision making
• Structuring
• Controlling and modelling

Organizations use information for a variety of purposes, complete the table by stating the information purpose and give another example for the scenarios listed.
    For example:

Scenario	Information purpose	Another example
A project manager needs to predict what resources will be needed to support the project	Forecasting	A company wants to gather information in order to predict sales over the next six months

Scenario	Information purpose	Another example
A small health food store wants to expand in the future and acquire new premises		
A leading clothes designer is considering what new collections to bring out next season		
An organization uses a set of measurements and calculations to design a new automated system		
New staff member information is used to identify seating plans in the new office		
Measurements are taken to determine the capacity of machinery loads and adjust them accordingly		
An employer needs to select a candidate for the job from the six interviewed		

Worksheet 18

Factors for planning

There are a number of ways to plan in an organization. The way in which you plan depends upon a number of factors. These factors evolve around a TROPIC cycle.

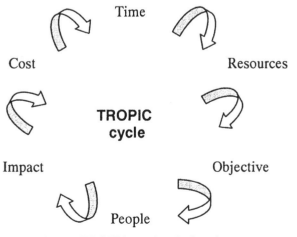

Figure 3.7 *TROPIC cycle of planning*

- **T**ime – how long is the plan
- **R**esources – what is going to support the plan
- **O**bjective – what is the purpose of the plan
- **P**eople – who is involved with the plan
- **I**mpact – who and what will the plan affect
- **C**ost – how much is it going to cost to implement the plan

1. Planning can take place over short, medium and long terms.
   (i) What length of time is associated with these terms?
   (ii) Identify a suitable plan that could be attributed to each term. For example, long-term planning might involve moving business into a new international market
2. Identify five resources that would be needed to support the successful implementation of a network system into an office.
3. State four objectives of setting up a company website.
4. Of what importance are the following people in a planning process?
   (i) Senior management
   (ii) Customers
   (iii) End users

Worksheet 19

Impact Analysis

One way to measure the impact of a new plan or project is to carry out an impact analysis, to see how, to what extent and whom the proposals will affect.

1. Using the example of a company that wants to install a computerized customer ordering system into a department that has very little or no IT skills, draw up an impact analysis to identify the effects of this proposal.

Cost is a major factor when deciding whether or not to go ahead with a new project plan. To justify the outlay a cost–benefit analysis can be drawn up to balance each associated cost with a benefit/s.

For example: the cost of buying a new PC for home use, the benefits of being able to do coursework and assignments at home, having the ability to learn new IT skills and do personal accounts.

2. Using the scenario of the company that wants to install a computerized customer ordering system into a department that has very little or no IT skills complete the given cost–benefit analysis

Cost	Benefit
1. Initial hardware outlay	
2. Customized software for customer ordering process	
3. Training of staff	
4. Maintenance and system management costs	
5. Expansion of office to accommodate the new computers	

3.11   Types of IS

Worksheet 20

General information systems

A variety of information systems are used in organizations. Some of these can be categorized as being general and others are more specific. General information systems are based on the use of applications software such as databases and spreadsheets to provide storage, modelling and prediction facilities. Specific information systems include: strategic level systems, management level systems, knowledge level systems and operational level systems.

1. Using general information system software design create a holiday booking system which can store information about customer bookings, flights and resort information. The system should include at least three tables and a form (relating tables is an option). Suggested screens could include:
    (i)   Customer details (customer number, name, address, contact details, preferences etc.)
    (ii) Flight information (e.g. flight code, airport, destination, flight time etc.)
    (iii) Resort information (resort code, location, accommodation type, star rating, meal basis, price etc.)
2. Design a sales forecast system that will automatically calculate:
    (i)   Profit, loss and break-even for the sales of the following computer items
    (ii) Which computer item is the most profitable at the end of this period?
    (iii) Which computer item is the least profitable at the end of this period?
    (iv) If the number of items sold increased by 20% per month (based on April's figures) how long would it take for the items that are making a loss to break even?

Computer item	Fixed costs	Cost of production (per item)	Selling price (per item)
Scanner	£2650	£28.50	£98.50
Printer Type A	£1450	£15.00	£62.00
Type B	£1450	£18.00	£71.00
Type C	£2450	£18.50	£76.00
External modem	£3400	£22.00	£87.50
Webcam	£1850	£15.50	£46.00

Sales orders for the next three months are as follows:

Months	Computer item	Sales
January	Scanner	10
	Printer Type A	12
	Printer Type B	21
	Printer Type C	15
	External modem	24
	Webcam	8
February	Scanner	5
	Printer Type A	8
	Printer Type B	3
	Printer Type C	9
	External modem	12
	Webcam	7
March	Scanner	8
	Printer Type A	10
	Printer Type B	1
	Printer Type C	11
	External modem	21
	Webcam	3
April	Scanner	7
	Printer Type A	22
	Printer Type B	1
	Printer Type C	14
	External modem	10
	Webcam	4

3. Template designs for both systems should be clear and fully functional (data is not required for the first system but can be added to make the system more comprehensive).

3.12 Specific information systems

Worksheet 21

Specific information systems types

Specific information systems are set up to provide support to a given level of management or organizational function. There are four categories of specific information system including:

- Strategic level systems
- Management level systems
- Knowledge level systems
- Operational level systems

1. Describe in detail what each of the four categories of specific information system provide in terms of functions and support.
2. What levels of an organization do each of the four categories support (i.e. strategic, tactical or operational) and why?
3. What types of information system fall within each of the four categories shown in Figure 3.8?

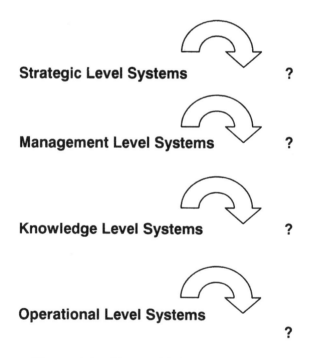

Figure 3.8 *Types of information system*

4. Provide a detailed overview of the following system types:
   (i) Expert system
   (ii) Decision support system
   (iii) Process control system

3.13 IS development bespoke v. Off-the-shelf software

Worksheet 22

Benefits and limitations of software types

Software can be categorized in many different ways, for example utility, operating system or applications software. Within this categorization another distinction to be made is whether or not the software has been bought off the shelf or if it has been designed to order, or 'bespoke'.

1. Examine and use three pieces of off-the shelf general applications software and carry out the following:
    (i)   Rate each piece of software out of 10 in terms of usability, screen design, help facilities, menu options, practicalities such as saving, updating and deleting information
    (ii) What were the benefits of using the applications software?
    (iii) What were the limitations of using the applications software?
2. What would you change in each piece of applications software to make it better?
3. If you were designing a piece of applications software what would be your primary objectives?
4. What would you say are the benefits of having bespoke software?
5. What are the limitations of having bespoke software?
6   Identify four situations where you would use (i) applications software and (ii) bespoke software.

Worksheet 23

Software options

A retail organization has decided to update its manual processing systems to a new computerized system that will gradually be implemented over the next three months. The departments that will be involved with this include:

* Sales and marketing
* Despatch
* Finance
* Personnel
* Operations

It is hoped that the new system will increase efficiency, reduce overheads in the long term and make the company more competitive in the short term.

1. Identify typical activities that each functional department would be involved with.
2. Describe what sort of software each department might need.
3. State whether or not bespoke software would be needed and if so what the benefits of having this would bring.
4. Identify and justify the typical costs involved with the software requirements on a stand-alone computer basis (the company will not be looking at a networked system for at least a year).
5. Identify any other issues and costs that would need to be considered with the software options you have chosen.

3.14   Systems lifecycle

Worksheet 24

The need for systems analysis

There are two main reasons for carrying out systems analysis: a cost reason and/or an efficiency reason.

1.   Give five examples of companies that would require a systems analyst for:
     (i)   Cost reason
     (ii) Efficiency reason
2.   Complete the table, giving examples of both cost and efficiency problems associated with each scenario that would require a systems analyst.

Scenario	Cost reason	Efficiency reason
Launching a product onto a new European market		
Upgrading from a manual to a computerized system		
Considering a new system to transport perishable goods quicker to retailers		
Identifying whether a store branch should be closed		
Deciding whether to set up online purchasing facilities		

Worksheet 25

Feasibility

1.   What is the importance of carrying out a feasibility study?
2.   What are the four recognized methods of collecting information during the feasibility stage? For each method identify:
     (i)   The type of information that would be collected, e.g. qualitative or quantitative
     (ii) How the information collected could assist with your further investigation
     (iii) Benefits and limitations of each method
3.   How could the information that has been collected during this stage be stored, e.g. what recognized templates or models are based on this information in later stages of systems analysis?
4.   What recognized steps are applicable to the feasibility stage?
5.   What is the final output from a feasibility study?

Worksheet 26

Data flow modelling

To produce a data flow diagram, four specific tools are used, these are:

- External entities
- Process boxes
- Data flows
- Data stores

1. Describe the function of each of the tools used to produce a data flow diagram.
2. When you produce a data flow diagram there are certain rules governing which tools can be linked to which, describe these rules.
3. For the following systems identify three examples of each of the data flow diagram tools.

Tool	Flight system	Library system	College system
External entity 1 2 3			
Process box 1 2 3			
Data flow 1 2 3			
Data store 1 2 3			

3.15 Case study – Bobs and Bits Hardware

Jonathon Primble, the manager of 'Bobs and Bits Hardware' called in a systems analyst about five weeks ago to examine how he could make his operations more cost effective, profitable and efficient. Last week, however, the systems analyst was taken seriously ill and could not complete the task. To date the information that has been gathered from an initial feasibility study is given below.

Feasibility findings

Personnel: Jonathon Primble   Manager
Joyce Carpenter   Sales
Graham Stooge   Stock control and supplies
Margaret Fischer   General assistant

Day-to-day operations and tasks

Each day Jonathon opens the store at 8:30. After switching everything on and opening the shutters, he fills the till and runs off a till check to confirm how much there is as an opening balance. The other staff members arrive between 8:30 and 9:00. Joyce is involved with sales and deals directly with the shop's larger suppliers and outlets. Each day she rings each corporate customer enquiring if they want any stock for the day. Once she has contacted them she makes a note on her corporate customer list by ticking against their name and details.

If stock is required she writes down the details on the stock ordering form and passes it over to Graham who is in charge of the stock and supplies. Jonathon ensures that items that have been requested are available in the storeroom, if they are he will get the stock order together. As items are taken from the storeroom the stock sheet is adjusted. When items reach a minimum level (varies for each stock item) a new stock request form is filled in, the top copy is sent to the suppliers and the bottom copy is retained and filed in the stock cabinet.

Margaret's role is to deal with customers as they come into the shop. She serves them, takes the payment and gives them an electronic till receipt and a written stock receipt detailing the items purchased (top copy is given to the customer and the bottom copy is stored in the receipts file).

At the end of the day Jonathon takes the money and cheques etc. from the till to the bank and locks up ready for the next day.

Worksheet 27

Preparing data flow diagrams

1. Identify any problems that exist with the Bobs and Bits Hardware system.
2. How could these problems be solved?
3. Prepare a context diagram and a Level 1 DFD for this system.

Worksheet 28

Designing a system

Jonathon Primble of Bobs and Bits Hardware was very impressed with your work that you did in the early stages of the systems analysis project and has asked if you will continue as the other analyst is still off sick. Jonathon has asked if you can put forward a proposal to include some screen designs to move the system over to a computer.

1. Put forward a proposal which is fully justified in terms of meeting the needs of Bobs and Bits Hardware.
2. Design and develop a suitable system which would meet the needs of Bobs and Bits Hardware.
3. Jonathon is concerned about how much the new system will cost and how long it will take to install and transfer over existing data. Prepare an implementation plan covering all of these elements to put Jonathon's mind at rest.

3.16   Health and safety

Worksheet 29

Health and safety quiz

There are a number of health, safety and security issues to consider when working with computer and information systems. Some of these issues are amplified by the enforcement of certain legislation that protects people for which information has been stored and can be accessed electronically.

1. Identify at least three pieces of legislation that protect users under the framework of 'Health and Safety at Work'.
2. Identify four ways in which you can make sure that data is kept secure on a computer.
3. What penalties can be imposed on people who hack into systems and gain unauthorized access?
4. Identify four things that can improve the general health and safety within an office environment.
5. What is meant by the following terms:
   (i)   Virus?
   (ii) Encryption?
6. As a home user what could you do to ensure that information accessed via the internet is secure?
7. Is it safe to buy products and services online; Discuss.
8. What does the Data Protection Act do, and who does it protect?
9. If you were working at a computer workstation all day every day, what should you do to ensure your own health and safety?
10. Do you believe that the growth in e-commerce has had a positive or a negative effect of health and safety. Discuss.

3.17 Business Information Systems Unit test

Worksheet 30

Unit test

This test can be used to assess how much of an overall understanding students have of business information systems. It is recommended that the test is set under exam conditions over a period of 1½ hours.

Questions

Operational systems are used for the tasks involved in the daily running of a business.

1a. State what is meant by the following acronyms:
   (i)   TPS
   (ii) EPOS
   (iii) OAS   **(3 Marks)**
1b. Provide TWO examples of a transaction. **(2 Marks)**
2a. Identify the three levels of decision making. **(3 Marks)**
2b. State the appropriate level of decision making for TPS. **(1 Mark)**

Organizations can be structured differently depending upon their size, resources and function.

3a. Identify and draw TWO organizational structure. **(4 Marks)**
3b. Give ONE example for each structure of an organization that would fit
    that structure. **(2 Marks)**
4a. Identify FIVE functional departments in a typical organization and
    provide a description of what each department would do. **(10 Marks)**
4b. List TWO ways in which multinational companies can be organized. **(2 Marks)**
5a. Examine the following organization chart and list any problems that
    might occur as a result of this structure: **(7 Marks)**

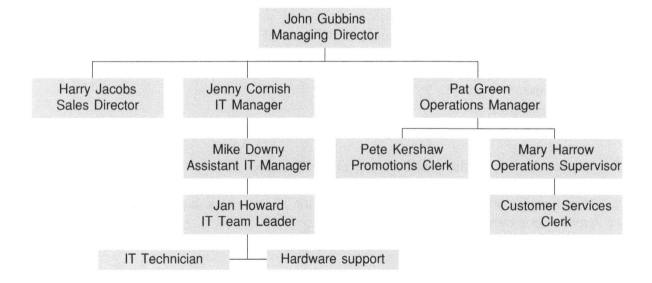

5b. Draw a revised organizational chart which rectifies these problems.   **(5 Marks)**

6. Identify the ways in which communication can flow through an organization.   **(3 Marks)**

7. What is the missing element on the communications diagram?   **(1 Mark)**

Data processing systems are systems that have been set up to process, manipulate and store volumes of data.

8. Give THREE examples of data processing systems.   **(3 Marks)**

9. Identify FOUR advantages of processing data automatically.   **(4 Marks)**

There are a number of different ways of representing the movement of information within an organization.

10. What name is given to this particular diagram?   **(1 Mark)**

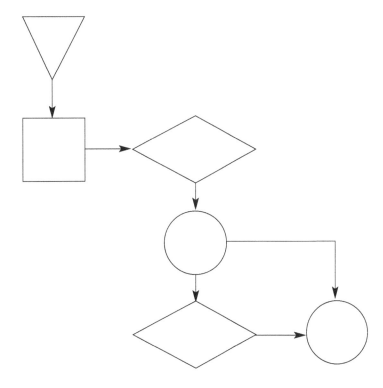

Health and safety rules and regulations are vital to all organizations to protect the end users from hazards that might occur in the workplace.

11. Identify TWO things that a computer user should do when working at a PC for long periods of time.   **(2 Marks)**

Rules and regulations have also been put in place to protect consumers from companies that hold information about them.

12a. Identify ONE Act that supports a consumer's right to have information
kept secure. **(1 Mark)**

12b. What measures can an organization take to ensure that consumer
information is kept secure? **(3 Marks)**

Business information systems are pivotal to any organization. Drawing upon your knowledge:

13. Describe in detail why business information systems are important to
organizations. **(5 Marks)**

14. How can business information systems improve communications in an
organization. **(5 Marks)**

15a. Identify TWO types of data processing methods. **(2 Marks)**

15b. State ONE difference between the data processing methods identified. **(1 Mark)**

PAPER TOTAL **(70 Marks)**

Unit 4 Software development

4.1 Fault finding in Pascal programs

Worksheet 1

Pascal

The following Pascal program has at least one syntax or logical error on each line. Identify each one.

```pascal
program worksht one (input;output);

var;  x,y,z:integer;
      earnings,tax,x:real;
      st,for,name:string

begin
      write("What is your name ");
      readln(yourname);
      writeln('Hello ',name, ", how much do you earn each year? );
      readline(earnings);
      tax=earning*0.22;
      writeline(You will pay', tax:0:3,' pounds tax');

end;
```

Write your answers as Pascal comments near each fault.

Worksheet 2

Pascal

The following Pascal program has at least five syntax or logical errors.
Identify each one.

```pascal
program 2ndversion_of_table;
var counter:integer;
      i:real;

function getnumber(lower,upper:integer):integer;
var inputstr:string;
      errorcode:integer;
      x:integer;

begin
      while
            write('Type a value between ',lower,' and ',
upper,' inclusive');
            readln(inputstr);
            val(inputstr,x,errorcode);
      until ((errorcode=0) and (x>=lower) and (x<=upper));

end;

begin {start of main program block }

      counter:=getnumber(2,12);
      for i:=1 to 12 do;
                  writeln(i,' times ',counter,' = ',i*counter);
end.
```

Write your answers as Pascal comments near each fault.

Worksheet 3

Pascal

The following Pascal program has many syntax or logical errors.
Identify each one.

```
program area;

var a,b:real;
     x,y:float;


procedure area(a,b:real);
begin
     var area,a,b:real;
     area=a*b;
end.

begin
     print("Program to print area of rectangle");
     print("What is the length of side A? ");
     readln(a);
     print('What is the length of side B?);
     readln(b);
     area(a,b);
     print(The area is ',area:0:2);
end;
```

Write your answers as Pascal comments near each fault.

Worksheet 4

Pascal

The following Pascal program has many syntax or logical errors.
Identify each one.

```
program conversion(output,input);
uses crt;

function miles2km(x:integer);
start
      milestokm:=x/1.609344;
end;

begin
      write(Program to convery miles to kilometres');
      write('What is the distance in miles? ')
      readln(miles);
      writeln(miles,' equals ',miles2km(miles),' kilometres);
      clrscr;
end.
```

Write your answers as Pascal comments near each fault.

Worksheet 5

Pascal

This is a repeat of Program 4.23 from the book but with a single error introduced. Find the error by using appropriate *test data.*

```pascal
program prog4_23;

function titlecase(name:string):string;
{ returns string with first letter=capitals }

var   p,i:integer;
         ch:string;
         done:boolean;
         convert_to_caps:boolean;
         outputstr:string;

begin

     name:=lowercase(name); { first ensure everything is lowercase }

     done:=false;
     repeat {a loop to ensure that any leading spaces are removed }

          if copy(name,1,1)=' ' { if first char is a space }
               then
                    { get rest of string only }
                    name:=copy(name,2,length(name)-1)
          else
                    { assign boolean to end the loop }
                    done:=true;
     until done;

     convert_to_caps:=true; { initialise boolean for use in for loop }
     outputstr:=''; { initialise string to null
                         ready to build output string}

     for i:=1 to length(name) do; {look at each character in string }
          begin
               ch:=copy(name,i,1); { get one char at a time }
               if convert_to_caps=true
                    then
                    begin
                         { add upper case char to output string }
                         outputstr:=concat(outputstr,upcase(ch));
                         { set boolean ready for next time}
                         convert_to_caps:=false;
```

```
                            end
                            else
                            begin
                                    { add lower case char to output string }
                                    outputstr:=concat(outputstr,ch);
                            end;

                        { see if next char must be capital }
                        if ch=' ' then convert_to_caps:=true;
                end;
        titlecase:=outputstr; { assign finished string to the function }
end;


begin {start of main program block }
      {simply to test function output}

writeln(titlecase('this is a test string'));
end.
```

Worksheet 6

Pascal

Paint estimator 1

Write a program that estimates the amount of paint required to decorate the walls of a plain rectangular room. Your program should ask the user for the room's dimensions and the paint coverage per litre, then calculate the amount of paint required for a single coat of paint.

- Allow a yes/no question to ask if the ceiling is to be included in the calculation.
- Assume a plain rectangular room.
- Make no allowance for windows or doors.
- Do not use any data input validation or error checking.
- For a room with painted walls, emulsion paint covers X m^2 per litre. Use a coverage of 15 m^2 per litre for program testing but your program should allow any value.

The calculation is ((area of wall A*2) + (area of wall B*2))/paint coverage. If the ceiling is to be painted, use the wall dimensions to calculate ceiling size then round up to the nearest litre. Only round up total wall + ceiling paint required at the end of the program to avoid two rounding errors.

Points to note

- Use stepwise refinement to design your program.
- *Before* coding, design simple test data.
- Use sensible variable and procedure names.
- Use appropriate variable types.

- When rounding up the paint required, allow for a small error in coverage, i.e. if the amount of paint is only 2% or less above an integer number of litres, round it down. For example, suppose the paint required was 7.03 litres. 2% of 7 litres = 7*2/100 = 0.14. As 0.03 is below 0.14, round it down. Putting it another way, 7.14/7 = 1.02 meaning 7.14 is 2% above 7.

Worksheet 7

Pascal

Paint estimator 2

Extend your paint estimator program to calculate the paint costs.

A local paint shop sells paint with these prices. Extend the program to ask if white or a colour is required then ouput the paint cost. Only allow for the ceiling to be the same colour as the walls.

Size	Brilliant White	Colours
10 litre	£26.99	£39.99
5 litre	£13.49	£20.99
2½ litre	£9.29	£12.99
1 litre	£4.99	£6.49

You should use some integer arithmetic DIV and MOD to arrive at the correct number of tins to obtain the best price. For example, if a job required 12 litres of coloured paint, you could make 12 litres by

12 by 1 litre tins =	£77.88
2 by 5 litre tins plus 2 by 1 litre tins =	£54.96
4 by 2½ tins plus 2 by 1 litre tins =	£64.94

Clearly, the best price is £54.96.

Worksheet 8

Pascal

Paint estimator 3

Rewrite the paint estimator program to use procedures.
Write a brief answer to the questions:

1. What advantage does the use of procedures give over a single piece of program code regarding:
    (i)   The user?
    (ii) The programmer?
    (iii) The programmer's employer?
2. What disadvantage occurs when using only global variables?

Worksheet 9

Pascal

Paint estimator 4

Extend your paint estimator program to provide data input validation.
Decide on sensible minimum and maximum data inputs and formats, exclude any data outside these limits.

Allow the user to press the ENTER key *only* if:

- a default value of 15 m^2 is to be used for the paint coverage
- the ceiling is to be painted – assume an answer of no

Modify your test data before recoding as it will now have to include values that test the data validation.

Provide a short user guide and explain any limitations that your program may have.

Unit 5 Communication technology

5.1 Internet, the main search engines

Some search engines have information organized by humans (also called a directory), others have information organized by computers. The second kind uses software (called a spider, robot or crawler) to look at each page on a website, extract the information and build an index. It is this index you search when using the search engine. Some people make a clear distinction between a directory and a search engine. Currently, the situation is not clear cut as many 'search engines' in fact use both methods.

The performance of a search engine depends critically on how well these indexes are built. It is also very important to remember that the whole business of search engines is in a state of constant change. Companies buy each other, change systems, indexes etc. Some rely on other people's information. There is no such thing as a static search engine!

Table 5.1 lists some of the main search engines.

Table 5.1 *Main search engines*

AllTheWeb.com (FAST Search) http://www.alltheweb.com	One of the largest indexes on the web.
AltaVista http://www.altavista.com	One of the oldest crawler-based search engines on the web, it also has a large index of web pages and a wide range of searching commands.
AOL Search http://search.aol.com/	Uses the index from Open Directory and Inktomi and offers a different service to members and non-members.
Ask Jeeves http://www.askjeeves.com	Ask Jeeves is a human-powered search engine that introduced the idea of plain language search strings.
Direct Hit http://www.directhit.com	Direct Hit uses its own 'popularity engine' that depends on how many times a site is viewed to judge its ranking. This idea is not always successful as the less popular sites do not get a chance to rise, so popular ones remain popular. Direct Hit is owned by Ask Jeeves.
Google http://www.google.com	Google makes use of 'link analysis' as a way to rank pages. The more links to and from a page, the higher the ranking. They also provide search results to other search engines such as Yahoo.

HotBot http://www.hotbot.com	Much of the time, HotBot's results come from Direct Hit but other results come from Inktomi. Hotbot is owned by Lycos.
Inktomi http://www.inktomi.com	You cannot query the Inktomi index itself, it is only available through Inktomi's partners. Some 'search engines' simply relay what is found in the Inktomi index.
LookSmart http://www.looksmart.com	LookSmart is a human-compiled directory of websites but when a search fails, further results are provided by Inktomi.
Lycos http://www.lycos.com	Lycos uses a human developed directory similar to Yahoo and its main results come from AllTheWeb.com and Open Directory.
MSN Search http://search.msn.com	Microsoft's MSN is powered by LookSmart with other results from Inktomi and Direct Hit.
Netscape Search http://search.netscape.com	Netscape Search's results are from Open Directory and Netscape's own index. Other results are from Google.
Open Directory http://dmoz.org/	Open Directory uses an index built by volunteers. It is owned by Netscape (who are owned by AOL).
Teoma http://www.teoma.com	A new search engine, launched in April 2002, that claims to be better than Google.
Yahoo http://www.yahoo.com	Yahoo is a human-compiled index but uses information from Google.

5.2   Effective searching of the internet

Some subjects are hard to find as the words used in a search engine often lead to many different subjects.

The task here is to find specific information and to eliminate all the non-relevant information. There is no 'answer' as such, you either find what you want or not.

Example – the ingregients of chocolate powder

If you want to know what is in a chocolate powder, you might use the search string 'chocolate ingredients'.

Using this string, Google gave 417 000 'hits' and Lycos gave 7977. None of the hits at the top of the Google list were useful as they only contained the words 'chocolate' and 'ingredients'. One hit referred to a Parliamentary debate! Lycos gave fewer hits as the index is human organized so 'chocolate ingredients' is looked upon as a *subject*, not just *keywords*.

The key is to use whatever search engine/directory best suits your purpose. If you are searching for a 'subject', use a *directory*, if you are searching for specific information that can be isolated with a keyword or two, use a *search engine*. If necessary, use both.

To narrow the results, you can use the 'advanced' searches, though there is nothing advanced about them!

In Google Advanced, these words were entered:

with all the words	chocolate
with at least one of the words	ingredient ingredients components substance analysis
without the words	cake biscuit cookie sponge

This simply results in the search string:

chocolate ingredient OR ingredients OR components OR substance OR analysis – cake – biscuit – cookie – sponge.

where the OR operator is self-explanatory and the '–' operator means leave out pages with this word. (You can type such a search string directly into most search engines. Their rules vary a little but most use + to include a word a – to exclude it.)

This search string gave a link that contained this information:

sugar, modified palm kernel oil, modified milk ingredients, soya, lecithin, coco powder, artificial colour, vanilla and vanillin.

This is useful as it gives more keywords that you may not have thought of that should be used to narrow the search. Using these new keywords yielded a link that contained these ingredients:

cocoa processed with alkali, reduced minerals whey, maltodextrin, soy lecithin, tricalcium phosphate, salt, aspartame (non-nutritive sweetener), acesulfame potassium (non-nutritive sweetener), artificial and natural flavors.

The key point to note is that search engines or directories are simply an aid to searching, you must work at it to get the best results.

Can you trust the answers you find?

That depends on who published the information. The problem is no different when considering the internet, magazines or books.

If the site owner is a private individual, the information may be correct but it is not likely to have been checked with great rigour. If it tallies with similar sites, the chances are good that it is correct, but on the other hand, who copied who?

If it is an official site, the information will generally be an accurate reflection of the opinion of the site owners. Chocolate manufacturers will tell you chocolate is good for you, governments will tell you they serve you to the highest standards. Make your own judgement.

Academic sites often contain information from a more 'free thinking' group of people so will show a wide range of opinion. Specific scientific, technological or historical information is usually accurate, political views may be varied. Again, make your own judgement.

Worksheet 1

Searching the internet

Find out the following specific information. You are expected to use preliminary results to learn more about the subject, then use this new knowledge to search in more detail. You are not expected to know anything about the subject matter when you start.

Example

What is the name of the person who designed the world's first effective computer that was used to decode German military messages during the Second World War?

First results yielded the name of the computer, *Colossus*. It was made to decrypt the German Lorenz (not Enigma) codes.

A new search using the word Colossus yielded the information: *Tommy Flowers*. A genius. See http://www.codesandciphers.org.uk/.

Your tasks

Find out the following specific information. Only give results that you would trust, i.e. if you find information that seems to tell you what you want but are not *sure* it is correct, ignore it.

1. What is the fastest current Intel Pentium microprocessor?
2. What is the maximum data capacity of a commercially produced DVD?
3. What is the fastest current micropocessor fitted in Apple G4 machines?
4. What are the main side effects of taking the drug aspirin over a long period?
5. How many people, soldiers and civilians from all countries, were killed in the Second World War? Especially in this case, how can you be sure the answer you get is correct?
6. A British Prime Minister is quoted as having said 'There are lies, damned lies and statistics'. Who was it?
7. It is said that a British Prime Minister once included in a speech the phrase 'Blood, sweat and tears'. Is it true or did he say something else?
8. Who wrote the famous poem 'If'?
9. How many standard floating point number formats are specified by the IEE?
10. Imagine you have just been loaned a 50 foot motor boat by a rich friend. It is moored at Southampton. Is it legal for you to take it alone across the channel to France?

Unit 6 Systems analysis

6.1 The systems analyst

Worksheet 1

The role of the systems analyst

A systems analyst has to demonstrate a range of skills and abilities in order to ensure that any project undertaken will be a success. The skills that are required can be broadly categorized into three areas:

- Interpersonal
- Technical
- Application

1. From your own understanding of the tasks that a systems analyst has to do and drawing upon resource information, describe in detail why each of the three skill areas are important.
2. What skill would you rate as being the most important and why?
3. Which skill do you think is the hardest to develop and why?
4. It does not matter how you obtain the information for a feasibility study, or how you come across to users as long as you get the information you want. Is this a fair statement?
5. In the early stages of a systems analysis project what would you class as being crucial to the success of the project?

6.2 The need for systems analysis

Worksheet 2

Why carry out systems analysis?

Systems analysis is carried out to meet two broad demands. These revolve around financial demands (cost, profit, loss, budgets etc.) and efficiency demands (how can we improve this, become more productive, reduce overheads etc.).

1. Complete the table below by identifying both financial and efficiency reasons for carrying out systems analysis in certain functional areas of an organization.

Department/Function	Cost reasons	Efficiency reasons
Sales, e.g.	Over the past six months sales have fallen by 22%, as a result profit is down by £16 000	The current methods of obtaining sales orders from customers is not as productive as it used to be
Marketing		
Finance		
Customer services		
IT		
Production		
Despatch		

6.3 Lifecycle

Worksheet 3

Systems analysis lifecycle

1. Complete the systems analysis lifecycle model.

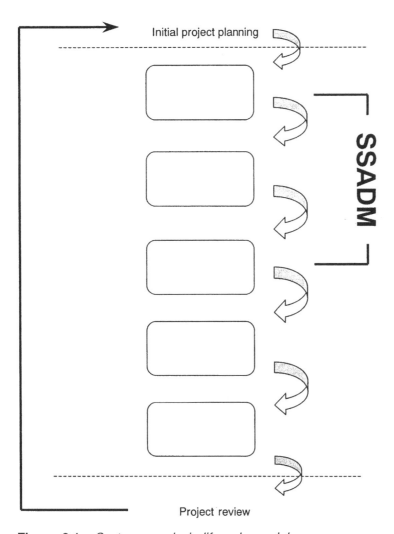

Figure 6.1 *Systems analysis lifecycle model*

2. SSADM is just one of many lifecycle models, others that exist include:
   (i)   Waterfall model
   (ii) Spiral model
   (iii) Rapid Applications Development Model (RADM)
   (iv) Dynamic Systems Development Model (DSDM)

Compare the lifecycle model of SSADM to one of the above models identifying similarities and differences in the stages of each.

6.4 Gathering information

Worksheet 4

Information sources

The most important stage of the systems analysis process is gathering information. Without information the project cannot develop and recommendations cannot be put forward.

1. Look at the following list of information sources and identify:
   (i)   What sort of information can be gathered
   (ii) How this information can assist in the systems analysis project – how it can be used
   (iii) Where the information could be found – the source
   (iv) Of what value is the information to the systems analysis project

   Information sources:

   - Organization chart
   - List of personnel – users of the system (outlining what job roles there are etc.)
   - Overview of the system
   - Resource list
   - Documentation used to support the system

2. What problems might occur when trying to obtain the following information:
   (i)   How the current system works
   (ii) What users do within the system – what their job role is
   (iii) What documents are used to support the system
   (iv) Identification of problems within the current system
3. You are in a situation where you have talked to two users within the system who work closely together and do almost identical tasks each day. However, both users have given two very different accounts of what tasks are done, how they are done, what is involved and the time it takes.
   (i)   What would you do with this conflicting information?
   (ii) How could you verify which version of information is more accurate?
   (iii) Why do you think the users would give different accounts of the same tasks?

Worksheet 5

Fact finding

The fact-finding stage of systems analysis falls within one of the early stages of the lifecycle known as 'feasibility'. The objective of this stage is to gather information using a variety of tools and techniques from a range of information sources in order to establish facts, verify problems and recommend solutions and changes.

There are a number of ways in which information can be gathered; however, the four traditional methods that are used frequently in systems analysis projects include:

- Observation
- Interviews
- Questionnaires
- Examination of documentation

1. Complete the table below by identifying the benefits and drawbacks of each method:

Method	Benefits	Drawbacks
Observation		
Interviews		
Questionnaires		
Investigation of documentation		

2. Under what circumstances would you use observation as a fact-finding technique?
3. Identify five documents that you could examine in a typical organization.

Worksheet 6

Starting a systems analysis exercise

One of the first tasks of a systems analysis project is to establish the area in which you would be working. This task is referred to as 'setting the systems boundary' or 'setting the scope' of the project.

1. Why is it important to define the scope or boundary of the systems analysis project?
2. Systems boundaries can be defined at various levels, for example:
    (i)   Geographically
    (ii) Departmentally
    (iii) By function
    (iv) By specialism
    Provide an organizational example for each of the above system boundary levels to show how they can be arranged.

3. What sort of information could you obtain by examining the following documents in the systems listed below and why would this information be of value?
   (i) Booking form for a holiday at a travel agents
   (ii) Enrolment form for a student at college
   (iii) Fault log sheet for an IT department
   (iv) Sales performance sheets of personnel within a electrical retail department
4. When you are interviewing a user of the system, what factors should be taken into consideration?

6.5 Fact finding

Worksheet 7

Using fact-finding tools

Questionnaires are one of the recognized tools used for collecting information in a systems analysis project.

1. Examine the sample questionnaire and identify which areas would need further investigation.

ID number: 013	System objective:	**Improve the despatch process to ensure faster customer deliveries**
Name: Graham Shay	Department: Despatch	Job title: Despatch Supervisor

Tasks undertaken each day:
- Check the day's delivery schedule
- Distribute deliveries to despatch team
- Liaise with sales about the next day's deliveries
- Complete and authorize the despatch sheet
- Arrange for overseas deliveries
- General supervision of the despatch department

Communicates with: despatch manager, despatch personnel (six delivery drivers, two clerks and two stock personnel)

Documents used:
- Delivery management sheet
- Stock despatch sheet
- Stock control sheet
- Sales deliveries schedule

Constraints and problems:
1. Too much repetition of information in the stock control and despatch sheets
2. Sometimes sales forget to send down the delivery details which holds up a customer's order being despatched
3. Overseas deliveries are expensive because only casual despatch companies are used depending on the location of the delivery
4. There is no co-ordination of deliveries, all despatches are designated each day on a random basis to each driver

Any other information:
Customers complain about the time it takes to deliver an item, especially when next day delivery orders are not met

2. How and where would you get this information from?
3. From the initial information given what proposals could you put forward based on the constraints and problems identified.

Worksheet 8

Using documentation - user catalogues

This task can be carried out in conjunction with Worksheet 9 - requirements catalogues.

The function of user catalogues is to capture information about users within the system by identifying what they do – job title and what their responsibilities are.

Look at the layout of the sample user catalogue.

User catalogue	
Job title	**Job role/responsibility**
Systems administrator	• Sets up passwords for users • In charge of PC consumables • Helps users with technical problems • Changes the backup tapes • Helps set up the company website
Network manager	• In charge of the network personnel in the IT department • Manages the network budget • Oversees network developments and projects • Involved with IT strategy planning

Using a similar structure carry out the following tasks:

1. Pick a suitable organization and set up a time to go in and talk to at least five users within that organization, preferably users who work within the same department. Iif you have difficulties in setting this up try the following:
   (i) Ask your parents, brother or sister if you could use their organization as an example
   (ii) If you have a part-time job base the exercise around that
   (iii) Write off to organizations explaining that this exercise is part of your National Diploma course and could they help
   (iv) Base the exercise on your school or college
2. Complete a user catalogue identifying what they do – job title and what their responsibilities are.

Worksheet 9

Using documentation – requirements catalogues

Requirements catalogues are used to track and record the requirements of users for the new system proposals. The requirements catalogue examines the functional and non-functional requirements of the new system and looks at the links between the requirements given from each user.

Tasks

1. Using the resources of the users selected for Worksheet 8 prepare a requirements catalogue. Identify any problems that they have with a system that they are currently working with and ask them how they would improve or change the system for the better.

Requirements catalogue for a new sales system			
Requirement ID: 117	Status: Essential	Source: Sales manager – Jack Hue	Users involved: Sales team, sales director, marketing department, despatch department
Functional requirement ☐ Non-functional requirement ☐		Requirement: To provide a more up-to-date and automatic sales system for customer orders	
Benefits to the system: • More efficient and faster ordering system • Easy and automatic access to customer records • Ability to combine marketing and sales data for future campaigns and targeting of customers • System linked directly to supplier and despatch department			
Non-functional considerations: Customer payment details should only be accessed by sales supervisor status or above			Action: Set up a password system
Comments:			
Dependency on other requirements: Requirement ID Status 116 Essential 115 Desirable 114 Desirable			
Proposed solution/s: 1. Install a computerized customer ordering system 2. Link the system into the existing marketing and despatch systems 3. Set up a password system 4. Set up an automatic ordering system to alert sales department of stock status and when to order			

6.6 Data flow modelling

Worksheet 10

Data flow modelling

Data flow modelling is an established SSADM tool that is used to examine the environment of the system under investigation through the use of data flow diagrams (DFDs) and associated descriptors. The descriptors are used to identify and establish the flow of information within a system, processing activities that take place and the types of storage mechanisms used.

Tasks

1. There are four tools that are used to prepare a data flow diagram as shown. Label and provide five examples for each tool.

(i)

Examples:

_____

_____

_____

_____

_____

(ii)

Examples:

_____

_____

_____

_____

(iii)

Examples:

_____

_____

_____

_____

_____

(iv)

Examples:

_____

_____

_____

_____

_____

Worksheet 11

Preparing a DFD

Before launching into a full-scale DFD it is better to prepare yourself by producing drafts and breakdowns of individual systems that operate within the completed DFD. The first place to start is to identify what is happening within the system by picking out what processes, data stores, data flows and external entities exist.

1. In pairs discuss with your partner an activity that you have recently done or an event that you attended. As you are talking your partner should prepare a table and record your activity under one of the four headings, e.g.:
   (i)   Process
   (ii) Data store
   (iii) External entity
   (iv) Data flow
2. Ensure that you take it in turns to speak and to record information.
3. When you have completed the first part of the exercise start to create a DFD for your partner's system by setting out the activities in process boxes as shown in the example.

Sample process DFD skeleton for booking a holiday:

4. Check that the processes identified are correct and then move onto adding the other tools – data stores, data flows and external entities.

Worksheet 12

Group design of a DFD

When you are preparing a DFD it sometimes helps to talk through your design to ensure that all of the information has been captured and is correct. One way of checking that a DFD design is correct is to design one within a group.

1. Within the class, identify if anybody has a part-time job or hobby at the weekends or in the evenings. Choose two people within the group who are willing to talk through what they do. As the information is being recalled each person within the group should record it using the headings of process, data store, data flow and external entity. Information recorded should include:
    (i)   What tasks they do
    (ii) Who they communicate with
    (iii) What documents they use
    (iv) Where information comes from and what happens to the information
2. Once they have finished each person within the group should have their own set of notes which will contribute to the DFD design.
3. Ask for volunteers to identify what they have recorded. As the information is being read out other members within the group should add to or question the information to ensure that the account is as accurate as possible.
4. Once everybody including the person who gave the account is satisfied a context diagram, level 1 and level 2 data flow diagram (if appropriate) can be drawn on the board representing that system.

6.7   Case study 1 – Fursham Wines

This case study has been designed to provide an interactive and hopefully enjoyable scenario to enable students to explore a wide range of learning outcomes as specified within the systems analysis unit.

The case study theme is continuous throughout the three assignments to provide as realistic an example as possible (a systems analysis project usually being based on a single organization or problem).

The three assignments are all modelled around the day-to-day activities of four company members and the managing director (no information has been supplied for this role).

Sarah – Marketing and Sales Manager
Richard – Operations and Distribution Manager
Pat – Accounts Manager
Ken – Customer Services Manager
Jasper – Managing Director

Accounts have been provided for each of the four main characters (these can be added to or adjusted as required).

The objective of this case study is to get people to take on each of the roles within the company. Students can then carry out their fact-finding investigation by giving them a questionnaire or setting up interviews. Role play is critical to the success of the case study. If resources and time are limited each person portraying the role could get together so that students could have an opportunity to get their information all at once. The information provided for each character can be added to as long as information given out is reasonably consistent with other accounts.

Documentation has not been provided but some descriptions have – again this can be expanded and documents produced if required.

Assignment time and assessments can be centre devised.

Fursham Wines Assignment 1

Fursham Wines

Fursham Wines is a new company that plans to market wine in the UK.

Initially the company will import, bottle and sell two wines from Australia, one red and one white. They have provisional names for both brands but have not yet made a final decision. Currently the company has a small office, no warehousing facilities as yet and imports have just started on a small scale.

The Australian wine will be imported in bulk containers through the port of Dover and transported by road. It is anticipated that this pattern will be repeated for subsequent products from other regions of the world.

Fursham's does not intend to sell cheap wines or wines from countries such as France and Germany as these trade sectors are already well supplied. All of their products will be sourced from less established wine producing countries and will be sold on the basis of their quality.

Wine sales generally are significantly higher in London and the South East than other areas of the country and it is anticipated that this is where the majority of Fursham's business will be. The major wine retailers with whom they hope to trade all have their head offices in this area.

The company has been founded by the following people who have all been friends since college.

EMPLOYEES
Sarah is the Marketing and Sales Manager within the organization, she deals with all of the advertising and selling of the products.

Richard is the Operations and Distribution Manager, he deals directly with all of the suppliers and distributors.

Pat is the Accounts Manager, she liaises with all members within the company, prepares all the invoices and other relevant documentation.

Ken is the Customer Service Manager, he deals with all enquiries and follow-up calls regarding any stock which has been sold.

Jasper is the Managing Director of Fursham Wines, he put up the majority of capital for the company, and oversees all day-to-day activities.

At present all of the day-to-day operations are done manually. However, plans to expand the company are in progress, and it is expected that another 15 people will come on board to assist the managers in their tasks.

Worksheet 13

Investigating systems

1. You need to carry out a feasibility study to establish exactly what goes on at Fursham Wines. Prepare templates and questions for at least two fact-finding methods of investigation.
2. Following the feasibility study, write a report to Jasper to cover the following points:
   (i) An explanation of the stages of systems analysis and what the investigation involved as Jasper knows nothing about this area
   (ii) The purpose of carrying out a feasibility study and what it has achieved
   (iii) A description of what each member of Fursham Wines does
   (iv) Identify who the members of Fursham Wines communicate with and the types of documents that they use
   (v) Identify the problems that exist
   (vi) Recommendations for overcoming these problems
3. Prepare a ten minute presentation based on your report findings which can be delivered to the members of Fursham Wines.

Fursham Wines Assignment 2

Fursham Wines

The founders of Fursham Wines were very impressed with the professional approach and advice given to them regarding their company.

They have expressed that they would like you to provide further analysis and investigation into their business. After providing in-depth detail about their existing system they would appreciate some proposals about a new system which could be used to solve some of the problems which were expressed at the first meeting.

Worksheet 14

Analysing systems

1. Produce a set of detailed 'business system options' and 'technical system options' for Fursham's based on your recommendations in the feasibility report.
2. Use data flow modelling techniques to represent the current system at Fursham's.
3. Produce an entity relationship diagram for the existing system.
4. Provide a summary of how the new system will affect each member within the company.

Fursham Wines Assignment 3

Fursham Wines

Following your investigation Jasper has decided to introduce computers into the company and update all existing operations into electronic formats.

Jasper has asked if you would complete your investigation by ensuring that the computers are successfully implemented. Concerns over the lack of IT skills of some of the company members have also prompted Jasper to ask if you could help with the design of a system which could be used by everybody within the company.

Worksheet 15

Designing systems

1. Investigate different methods of implementing a new system, identifying the benefits and limitations of each.
2. Based on your entity relationship diagram and other modelling tools used, design a suitable system which could be used for *one* of the following purposes:
   (i) Tracking and ordering of wines
   (ii) Supplier and customer information
   (iii) Sales and financial control
3. Demonstrate the use of your system within a formal environment.

Sarah – Marketing and Sales Manager

General role

1. Promotion of wines in London and the South East (primary) and the rest of England (secondary – due to less demand).
2. Promotion through:
   (i) Direct liaison with wholesalers
   (ii) Direct liaison with promotions company (Aztec) for the design of posters, leaflets and bottle labels
3. No selling through retail, wholesale only.
4. Setting up of taster sessions through wholesalers.
5. Checking of current market trends in terms of wine produce (who are the most popular country producers, who are Fursham's competitors?).
6. Liaising with other members of Fursham's to discuss pricing policies and marketing tactics on a month-to-month basis.

Documentation used

1. Graphic design template (GDT) – used for the design of posters and leaflets (details include date of promotion, type of promotion – description of purpose, graphic and slogan to accompany promotion).
2. Aztec costing sheet – provides a costing for GDT depending on the type of promotion, method of promotion, i.e. leaflet/poster and the size and type of graphic.
3. Taster sheet (TS) – used for setting up taster sessions (details include date, location, quantities of wine required, wine type, estimated loss of sales used in taster, estimated sales as a result of the taster session).
4. Market analysis sheet (MAS) – a tracking sheet which is set up to log wine produce and quality among suppliers each month.
5. Competitors' analysis sheet (CAS) – a tracking sheet which is set up to log sales of competitors' wines through distributors used by Fursham's on a monthly basis.
6. Sales report (SR) – illustrating the amount of wine sold by Fursham's each week.
7. Forecast report (FR) – used at the monthly meeting with Fursham Wine members to discuss pricing strategies and promotions for the next month.

Problems

1. Too much paperwork.
2. Too much time taken up at Aztec designs (is there any way of transferring the information about future designs more efficiently?).
3. A lot of the forms contain repeated information.
4. It would be nice to do the designs in-house (especially the graphics for the bottle labels).

Communication with

1. Distributors.
2. Aztec designs.
3. Suppliers.
4. Pat to go through costing and forecast sheets (all sheets are passed on to Pat for review and filing).
5. Richard often accompanies you on trips to see suppliers and also distributors (also a weekly meeting is set up with Richard to discuss the logistics of getting wine to distributors a lot quicker when they put through an order).
6. Ken to check that responses from distributors is favourable following a promotion.
7. Each month a meeting with all members including Jasper to discuss future strategies.

Pat – Accounts Manager

General role

1. Deals with all the invoices and paperwork received from:
   (i)   Suppliers
   (ii)   Distributors
   (iii) Aztec Designs
   (iv)   Trans-Europe
2. Prepares monthly profit and loss accounts and forecast reports.
3. Prepares all the monthly salaries for company members.

4. Prepares costing sheets for marketing activities, e.g. taster sessions.
5. Liaises with the building society to oversee any BACS payments from distributors.

Documentation used

1. Calendar (C) – for recording dates for payments.
2. Invoices (I).
3. Nominal ledger (NL) – for recording all sundry purchases and petty cash expenses.
4. Purchase ledger (PL) – for recording all wine purchased and any other associated supplies.
5. Sales ledger (SL) – for recording any wine sales from distributors.
6. Wage slips (WS).
7. P11 (P11) – for sending off details to Inland Revenue about national insurance, tax and pension.

Problems

1. Too much administration.
2. Everything is done manually.
3. Too time consuming.
4. If other members of Fursham Wines do not record payment details on the calendar payments aren't made.

Communication with

1. Distributors (invoices).
2. Trans-Europe (invoices).
3. Suppliers (invoices).
4. Aztec Designs (invoices).
5. Sarah to discuss pricing policies for marketing events and to collate costing and forecast information.
6. Jasper to discuss monthly profit and loss account.

Richard – Operations and Distribution Manager

General role

1. Negotiates the logistics of getting wine from the supplier to the UK.
2. Deals with all issues regarding freight and transportation.
3. Liaises with customs and excise at Dover.
4. Ensures that all wine orders are with distributors within two days.
5. Ensures that all wine is transferred from suppliers to Fursham Wines within seven days.

Documentation used

1. Supplier contracts (SC) – signed for a period of one year to ensure that suppliers only provide wine to Fursham Wines.
2. Trans-Europe contract (TEC) – signed for a period of one year to ensure that all road transportation is undertaken by Trans-Europe and that penalties will be incurred if wine is not delivered within seven days from Europe and within two days locally to distributors.

3. Travel log (TL) – all road travel is logged to state mileage, problems encountered (i.e. French blockades at the ports), driver, vehicle specification on a daily basis (drivers sign the log and give a copy back to Richard).
4. Customs and Excise Form 1b (CEFb) – filled in monthly to declare VAT on all wine imported into the country.
5. Operations report (OR) – issued to Jasper every month outlining progress throughout the month in terms of deliveries.
6. Goods received note (GRN) – checked off when wine is received from supplier.

Problems

1. Too much administration.
2. Sometimes it is hard to keep track on the drivers at Trans-Europe especially when they are bringing wine from the suppliers.
3. Transportation costs represent a major outlay for Fursham Wines because Trans-Europe only have large freight lorries which are not practical for short runs to distributors.
4. More communication with Sarah to arrange distributor meetings with her as she is rarely in the office at the same time.
5. When away from the office dealing with new suppliers it is difficult to record all the information on paper and also information is only fed back to the office on return.

Communication with

1. Distributors.
2. Trans-Europe.
3. Suppliers.
4. Sarah accompanies you on trips to see suppliers and also distributors (also a weekly meeting is set up with Sarah to discuss the logistics of getting wine to distributors a lot quicker when they put through an order).
5. Ken to check that distributors are happy with the order.
6. Each month a meeting with all members including Jasper to discuss future strategies.

Ken – Customer Services Manager

General role

1. Liaises with distributors to ensure that any stock sold is of an acceptable quality.
2. Visits distributors to check if a repeat order of wine is required.
3. Follows through any enquiries regarding poor quality wine produce.
4. Deals with all distributor enquiries and orders.

Documentation used

1. Distribution sheets (DS) – recording when and where wine supplies have gone.
2. Order form (OF) – to record orders taken at the office.
3. Distributor feedback sheet (DFS) – given to distributors on a monthly basis to ensure that the quality of stock delivered meets a required standard.

4. Distributor comment sheet (DCS) – general comments are recorded by Fursham Wines regarding customer feedback.
5. Follow-up form (FF) – taken to distributors following an order to try to generate further orders on-site.
6. After sales report (ASR) – provides a monthly review of customer satisfaction and follow-up orders.

Problems

1. Too much time taken up with distributor visits, more time needed in the office.
2. Some of the paperwork is unnecessary, a lot of the information received back from distributors could be recorded on a single form.
3. More communication required with Richard and Sarah to find out about new products and the status of current ones, e.g. is a bin line being discontinued?

Communication with

1. Distributors.
2. Suppliers.
3. Richard to discuss stock levels and orders.
4. Sarah to discuss forthcoming promotions so that this can be incorporated into the after-sales service.
5. Pat to pass on order details.

6.8   Case study 2 Part I – Store line Supermarkets

Store line Supermarkets is an established chain of supermarkets that are located across the country. Over the past six months the managing director of the chain Mr Thomas North has discovered that they are losing their proportion of the market to another competitor. Since the beginning of the year their market share has fallen from 16% to 12%.

Store line Supermarkets has 15 stores across the region all located in major towns or cities. The structure of the company is very hierarchical with the following lines of command which are generic across all branches.

Branch structure

Head Office structure

All of the functional departments are located at the head office, which has the following implications for each branch:

- All recruitment is done through head office for each of the branches which means that all the application forms have to be sent either by post or online (if the application was filled in online).
- All stock ordering is done through head office who have negotiated local supplier contracts for each of the branches.
- All of the promotions, for example 'buy one get one free', and all of the price reductions or special offers are filtered through from sales at head office.
- All salaries are paid via the finance department at head office.
- All deliveries and distribution is made through local suppliers in conjunction with head office instructions.

All of the branches communicate on a regular basis. Branches distribute surplus stock items to other branches if they are running low, to reduce supplier ordering costs.

East Anglia branch

Thomas North has asked for an investigation to take place based on a branch in East Anglia. You have been given some general information about the organizational structure; however, Mr North is keen for you to carry out your investigation at the branch. You have been given four weeks to carry out the analysis and feed back to Mr North.

Fact finding

Using a variety of fact-finding techniques you managed to collect the following information:

1. There are 150 employees at the branch
   (i)   Store Manager – Mr Johnston
   (ii)   Deputy Manager – Miss Keyton
   (iii)   Five store managers, assistant managers and five supervisors
   (iv)   Fifty full-time and part-time check-out staff
   (v)   Forty full-time and part-time shelf stackers
   (vi)   Ten stock clerks
   (vii)   Ten trolley personnel
   (viii) Three car park attendants
   (ix)   Twenty cleaners, gardeners, drivers and other store staff
2. Each of the store managers controls their own areas, with their own shelf stackers and stock personnel.

3. All stock ordering is batch processed overnight to head office on a daily basis by each of the store managers in consultation with the deputy branch manager.
4. All fresh produce is delivered on a daily basis. Non-perishable goods are delivered three times a week by local suppliers.
5. All bakery items are baked on-site each morning.

As each department seems to operate on an individual basis the first part of the investigation for week one will be focused on the fresh produce department.

Fresh produce system

Ann Prior and her assistant manager Mary Granger manage the fresh produce department. Within the department their supervisor, John Humphries, oversees six display/shelf stackers and four stock personnel.

After consultation with a range of employees the following account of day-to-day activities has been given.

Each day Ann holds a staff meeting within the department to provide information about new promotions, special discounts or stock display arrangements. Any information regarding new promotions comes through from head office. All information received regarding promotions etc. is filed in the branch promotions file. If any price adjustments need to be made that day the stock personnel are informed to check the daily stock sheets.

After the meeting the stock personnel liaise with the shelf/display personnel regarding new stock that needs to go out onto the shop floor. The information about new stock items and changes to stock items comes from the daily stock sheet. When new items have been put out or stock price adjustments are made they are crossed off the daily stock sheet.

Items that have arrived that day are delivered from the local fresh produce supplier, when the items come in the stock personnel check the daily stock sheet for quantities and authorizes the delivery. If items have not arrived or there is an error in the order a stock adjustment sheet is filled in, this is kept in the stock office. At the end of the day John will then inform Mary of the stock adjustments. Mary then sends a top copy of the adjustment sheet to head office and files a copy in the stock cabinet.

Information about stock items running low comes from the daily stock sheet. If an item is low a stock order form is completed. A top copy is sent to head office and a copy is filed in the stock cabinet. Orders should be made five days to the actual requirement of the stock, as head office then processes the information and contacts the local supplier. In an emergency local supplier information is held by Ann, who can ring direct to get items delivered. This, however, costs the company more money because a bulk order has not been placed. Authorization also has to be given by the operations manager at head office. In addition, information has to be filled in on the computerized stock request form which is e-mailed to head office each day, who send back confirmation.

Head office dictates that all documents filled in online also need to have a manual counterpart, one which is sent off and the other which is filed with the branch.

Problems with the system

1. Sometimes the network at head office is down which means that stock items are not received within five days.

2. The promotions are not always appropriate because of a lack of certain stock. Sometimes it works out that the stock for which the branch has a surplus is wasted because they cannot set their own promotions in-store.
3. The stock cabinet is filled to capacity and because everything is in date order it is difficult to collate information about certain stock items.
4. If there is an error in the stock delivery nothing can be supplied until the paperwork has been sent off to head office or authorization has been given, even if the supplier has the stock requirement on his lorry.
5. Too much paperwork.
6. Little communication with other departments.
7. Targets that are set by head office cannot always be met due to the stock ordering problem.
8. Some stock items that come in are not all bar-coded.

Worksheet 16

Fact-finding tasks

The following sets of tasks can be completed individually as a set under each heading, or set as a complete task series.

1. Write down a list of tasks that you would need to do during the feasibility study.
2. What methods could have been used to collect the information given under the fact-finding section? Why would these methods have been appropriate?
3. The information given under the fact-finding section is quite general, what other specific facts might you need?

Worksheet 17

Identification and resolution of problems

1. A number of problems have been identified within the fresh produce department. From the information given do any other problems exist that you can see?
2. What proposals could you put forward to overcome these problems?

Worksheet 18

System proposals and design

1. The majority of operations and stock procedures are dictated to the branches via head office. Would the majority of problems be reduced if each branch had more control and autonomy over their own daily procedures, justify your answer fully?

The branch has decided to pilot a new stock control system for the fresh produce department which may be implemented across all of the branches.

2. Design a suitable system using applications software which could meet the demands and overcome the current problems of the fresh produce department. The system should take into consideration the following:
   (i) The users of the system
   (ii) Input and output screen designs
   (iii) Types of information that will need to be included
   (iv) How the system could be integrated across all branches
3. Write a short report about the merits of your system outlining your decisions made in the design stages. Identify any problems with the design or any concerns that you would have if the system were to be implemented throughout the branches.

6.9 Logical data modelling

Worksheet 19

Logical data modelling

Logical data modelling provides a detailed graphical representation of the information used within a system and identifies the relationships that exist between data items.

1. For each of the following systems identify at least six appropriate entities:
   (i) College
   (ii) Cinema
   (iii) Bank
   (iv) Flight booking
2. Draw a logical data structure to represent each of these systems. What problems if any have occurred?
3. For each of the following systems complete either the entity name, relationship type or both.

(i)

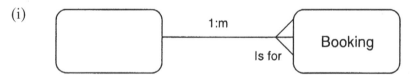

Figure 6.2 *Holiday system*

(ii)

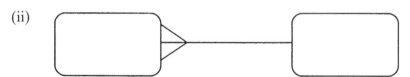

Figure 6.3 *Video/DVD rental system*

(iii)

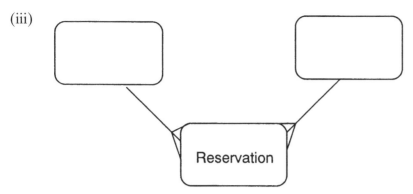

Figure 6.4 *Restaurant system*

(iv)

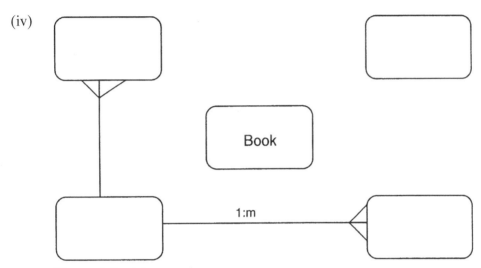

Figure 6.5 *Library system*

3. For the following entities, provide a complete attribute list with a suitable primary key:
   (i) Customer
   (ii) Student
   (iii) Patient
   (iv) Resort
   (v) Transaction
   (vi) Reservation
4. Why is it important to have a primary key for each entity?

6.10 Case study 2 Part II – Store line Supermarkets

The head office at Store line has decided to install computers in the fresh produce department of each of their branches. The investigation carried out did indeed raise many issues about the way in which stock processing and control is handled and the relationship between each branch and head office. As a result Store line have initiated a six month plan to ensure that each branch has adequate hardware and software provisions. The management team has set the following conditions for the new system implementation.

1. Each branch should be equipped with five computers in the fresh produce department at a cost of no more than £10 000.

2. The computers should have the capacity to be networked with other departments in the future.
3. One of the computers should be permanently linked to head office to ensure a continual download of offers, price changes and promotions.
4. The branch promotions file and daily stock and adjustment sheets should be computerized and information should be electronically transferred to head office when required.
5. Full system back-ups should be carried out overnight every night.
6. Ann Prior and Mary Granger would be the only personnel to have full access to all of the screens.
7   All personnel within the department should be capable of using the new system.

6.11 Business system options

Worksheet 20

Producing business system options

Business system options provides support and justification to the proposals given in the feasibility study. The main objective is to use the business system options to overcome and provide a solution for the requirements of the new system.

Store line Supermarkets have decided to go ahead with the new system implementation provided the conditions stipulated can be addressed.

1. Identify factors for consideration during the implementation of this new computerized system.
2. Based on the overall costing of £10 000 for five computers provide a full breakdown of tangible and intangible costs.
3. Produce a cost–benefit analysis based on your findings.
4. What impact will the new system have upon the personnel within the fresh produce department, produce an impact analysis to reflect this.
5. How will this new system address some of the problems identified in the earlier fact-finding scenario?

6.12 Technical system options

Worksheet 21

Technical system requirements

Technical system options (TSOs) examine a range of technical, developmental, organizational and functional aspects of the newly proposed system.

1. The requirements of the new system at Store line Supermarkets are quite broad. Carry out research and present a technical specification proposal for the computers, peripheral and software requirements for the fresh produce department.

2. There are a number of factors or constraints that can affect the implementation of a new system. Using the information given in the scenarios and from your general knowledge of system constraints identify at least four system constraints and state how these can be overcome.
3. What security issues will need to be addressed?

6.13 Systems design

Worksheet 22

Problems of a new system

Once new system proposals have been agreed, the next stage in the process is to design suitable user interfaces, input screens and menu systems. The introduction of a new system can cause many problems for an organization. These problems can include compatibility with existing systems, transference of data between systems and lack of understanding of the new system by current users.

1. Each member of personnel within the fresh produce department is required to use the new system. What measures can be taken to ensure that users understand the system and feel confident using it?
2. There are a number of ways that the new system can be implemented. Explore two implementation methods and select one of these options justifying your decision fully.
3. What measures can be taken after the system is installed to ensure that operations continue to run smoothly?

Worksheet 23

Input and output screen design

Store line Supermarkets have decided to use a combination of standard applications software and also bespoke software for their new systems. They have asked if you, as the original analyst, could put forward some draft screen designs for a stock control system.

1. Using suitable applications software design suitable input screens that could be used within the fresh produce department.
2. Identify three output screens that would be needed to support the requirements of the fresh produce department.

ASCII character set presented as control characters and text characters

The ASCII control characters, values 0–31

Dec	Hex	Keyboard	Binary	Description	
0	0	CTRL @	00000	NUL	Null Character
1	1	CTRL A	00001	SOH	Start of Heading
2	2	CTRL B	00010	STX	Start of Text
3	3	CTRL C	00011	ETX	End of Text
4	4	CTRL D	00100	EOT	End of Transmission
5	5	CTRL E	00101	ENQ	Enquiry
6	6	CTRL F	00110	ACK	Acknowledge
7	7	CTRL G	00111	BEL	Bell or beep
8	8	CTRL H	01000	BS	Back Space
9	9	CTRL I	01001	HT	Horizontal Tab
10	A	CTRL J	01010	LF	Line Feed
11	B	CTRL K	01011	VT	Vertical Tab
12	C	CTRL L	01100	FF	Form Feed
13	D	CTRL M	01101	CR	Carriage Return
14	E	CTRL N	01110	SO	Shift Out
15	F	CTRL O	01111	SI	Shift In
16	10	CTRL P	10000	DLE	Date Link Escape
17	11	CTRL Q	10001	DC1	Device Control 1
18	12	CTRL R	10010	DC2	Device Control 2
19	13	CTRL S	10011	DC3	Device Control 3
20	14	CTRL T	10100	DC4	Device Control 4
21	15	CTRL U	10101	NAK	Negative Acknowledge
22	16	CTRL V	10110	SYN	Synchronous Idle
23	17	CTRL W	10111	ETB	End of Transmission Block
24	18	CTRL X	11000	CAN	Cancel
25	19	CTRL Y	11001	EM	End Medium
26	1A	CTRL Z	11010	SUB	Substitute or EOF End Of File
27	1B		11011	ESC	Escape
28	1C		11100	FS	File Separator
29	1D		11101	GS	Group Separator
30	1E		11110	RS	Record Separator
31	1F		11111	US	Unit Separator

ASCII text characters

This table was produced using Excel 97 running on Microsoft Windows operating system. ASCII values 128 to 255 are not standard so other operating systems may yield different results. ASCII is based on the first 7 bits so affect values 0–127.

Dec	Hex	Binary	ASCII	Dec	Hex	Binary	ASCII	Dec	Hex	Binary	ASCII
32	20	100000		69	45	1000101	E	106	6A	1101010	j
33	21	100001	!	70	46	1000110	F	107	6B	1101011	k
34	22	100010	"	71	47	1000111	G	108	6C	1101100	l
35	23	100011	#	72	48	1001000	H	109	6D	1101101	m
36	24	100100	$	73	49	1001001	I	110	6E	1101110	n
37	25	100101	%	74	4A	1001010	J	111	6F	1101111	o
38	26	100110	&	75	4B	1001011	K	112	70	1110000	p
39	27	100111	'	76	4C	1001100	L	113	71	1110001	q
40	28	101000	(	77	4D	1001101	M	114	72	1110010	r
41	29	101001	)	78	4E	1001110	N	115	73	1110011	s
42	2A	101010	*	79	4F	1001111	O	116	74	1110100	t
43	2B	101011	+	80	50	1010000	P	117	75	1110101	u
44	2C	101100	,	81	51	1010001	Q	118	76	1110110	v
45	2D	101101	-	82	52	1010010	R	119	77	1110111	w
46	2E	101110	.	83	53	1010011	S	120	78	1111000	x
47	2F	101111	/	84	54	1010100	T	121	79	1111001	y
48	30	110000	0	85	55	1010101	U	122	7A	1111010	z
49	31	110001	1	86	56	1010110	V	123	7B	1111011	{
50	32	110010	2	87	57	1010111	W	124	7C	1111100		
51	33	110011	3	88	58	1011000	X	125	7D	1111101	}
52	34	110100	4	89	59	1011001	Y	126	7E	1111110	~
53	35	110101	5	90	5A	1011010	Z	127	7F	1111111	▓
54	36	110110	6	91	5B	1011011	[	128	80	10000000	€
55	37	110111	7	92	5C	1011100	\	129	81	10000001	□
56	38	111000	8	93	5D	1011101	]	130	82	10000010	,
57	39	111001	9	94	5E	1011110	^	131	83	10000011	*f*
58	3A	111010	:	95	5F	1011111	_	132	84	10000100	„
59	3B	111011	;	96	60	1100000	`	133	85	10000101	. . .
60	3C	11110	<	97	61	1100001	a	134	86	10000110	†
61	3D	111101	=	98	62	1100010	b	135	87	10000111	‡
62	3E	111110	>	99	63	1100011	c	136	88	10001000	^
63	3F	111111	?	100	64	1100100	d	137	89	10001001	‰
64	40	1000000	@	101	65	1100101	e	138	8A	10001010	Š
65	41	1000001	A	102	66	1100110	f	139	8B	10001011	‹
66	42	1000010	B	103	67	1100111	g	140	8C	10001100	Œ
67	43	1000011	C	104	68	1101000	h	141	8D	10001101	□
68	44	1000100	D	105	69	1101001	i	142	8E	10001110	

Dec	Hex	Binary	ASCII	Dec	Hex	Binary	ASCII	Dec	Hex	Binary	ASCII
143	8F	10001111	□	181	B5	10110101	µ	219	DB	11011011	Û
144	90	10010000	□	182	B6	10110110	¶	220	DC	11011100	Ü
145	91	10010001	'	183	B7	10110111	·	221	DD	11011101	Ý
146	92	10010010	'	184	B8	10111000	‚	222	DE	11011110	Þ
147	93	10010011	"	185	B9	10111001	^{1}	223	DF	11011111	ß
148	94	10010100	"	186	BA	10111010	º	224	E0	11100000	à
149	95	10010101	•	187	BB	10111011	»	225	E1	11100001	á
150	96	10010110	–	188	BC	10111100	¼	226	E2	11100010	â
151	97	10010111	—	189	BD	10111101	½	227	E3	11100011	ã
152	98	10011000	~	190	BE	10111110	¾	228	E4	11100100	ä
153	99	10011001	™	191	BF	10111111	¿	229	E5	11100101	å
154	9A	10011010	Š	192	C0	11000000	À	230	E6	11100110	æ
155	9B	10011011	›	193	C1	11000001	Á	231	E7	11100111	ç
156	9C	10011100	œ	194	C2	11000010	Â	232	E8	11101000	è
157	9D	10011101	□	195	C3	11000011	Ã	233	E9	11101001	é
158	9E	10011110		196	C4	11000100	Ä	234	EA	11101010	ê
159	9F	10011111	Ÿ	197	C5	11000101	Å	235	EB	11101011	ë
160	A0	10100000		198	C6	11000110	Æ	236	EC	11101100	ì
161	A1	10100001	¡	199	C7	11000111	Ç	237	ED	11101101	í
162	A2	10100010	¢	200	C8	11001000	È	238	EE	11101110	î
163	A3	10100011	£	201	C9	11001001	É	239	EF	11101111	ï
164	A4	10100100	€	202	CA	11001010	Ê	240	F0	11110000	ð
165	A5	10100101	¥	203	CB	11001011	Ë	241	F1	11110001	ñ
166	A6	10100110	¦	204	CC	11001100	Ì	242	F2	11110010	ò
167	A7	10100111	§	205	CD	11001101	Í	243	F3	11110011	ó
168	A8	10101000	¨	206	CE	11001110	Î	244	F4	11110100	ô
169	A9	10101001	©	207	CF	11001111	Ï	245	F5	11110101	õ
170	AA	10101010	ª	208	D0	11010000	Ð	246	F6	11110110	ö
171	AB	10101011	«	209	D1	11010001	Ñ	247	F7	11110111	÷
172	AC	10101100	¬	210	D2	11010010	Ò	248	F8	11111000	ø
173	AD	10101101		211	D3	11010011	Ó	249	F9	11111001	ù
174	AE	10101110	®	212	D4	11010100	Ô	250	FA	11111010	ú
175	AF	10101111	¯	213	D5	11010101	Õ	251	FB	11111011	
176	B0	10110000	°	214	D6	11010110	Ö	252	FC	11111100	ü
177	B1	10110001	±	215	D7	11010111	×	253	FD	11111101	ý
178	B2	10110010	^{2}	216	D8	11011000	Ø	254	FE	11111110	þ
179	B3	10110011	^{3}	217	D9	11011001	Ù	255	FF	11111111	ÿ
180	B4	10110100	´	218	DA	11011010	Ú				

DEBUG is a very odd program! It can be used to look into RAM directly or into any file, it can assemble or unassemble files, it can run programs and be used to change as little as 1 bit.

It is started at the DOS prompt by

C:\>DEBUG yourfile.txt

and all you get is a '-' character!

This '-' character is the input prompt. Possible commands are shown below.

```
assemble     A  [address]
compare      C  range address
dump         D  [range]
enter        E  address [list]
fill         F  range list
go           G  [=address]  [addresses]
hex          H  value1  value2
input        I  port
load         L  [address]  [drive]  [firstsector]  [number]
move         M  range address
name         N  [pathname]  [arglist]
output       O  port byte
proceed      P  [=address]  [number]
quit         Q
register     R  [register]
search       S  range list
trace        T  [=address]  [value]
unassemble   U  [range]
write        W  [address]  [drive]  [firstsector]  [number]
allocate expanded memory         XA  [#pages]
deallocate expanded memory       XD  [handle]
map expanded memory pages        XM  [Lpage]  [Ppage]  [handle]
display expanded memory status   XS
```

For the use that DEBUG is put to here, to produce a hex dump, the only commands needed are D for Dump and Q for Quit.

To produce a hex dump of a file called TEST.DAT simply issue the command

C:\> DEBUG TEST.DAT

at the command prompt then use the command D (then enter). You can 'dump' from any address and any size, the command D 100 1000 dumps from address 100 and dumps 1000 bytes, both numbers in hex.

If you need the hex dump in a file, you should first prepare a command file that contains just the data below. Assume this command file is called CMD.DAT, it is made using the commands at the DOS prompt

C:\> COPY CON CMD.DAT
D
Q
CTRL Z

where CTRL Z means press the CTRL and Z keys together.

You can now produce a hex dump in a file by using I/O redirection as in the command:

C:\> DEBUG TEST.DAT < CMD.DAT > TESTHEXDUMP.TXT

The < character means take input from your command file called CMD.DAT and the > character means place or re-direct the output into whatever file name you provide, in this case TESTHEXDUMP.TXT.

If you need to dump more than 100 hex bytes, the file CMD.DAT will have to contain

D 100 200
Q
CTRL Z

where the 200 refers to 200 hex bytes in length. If you need to quote a length in the D command, the starting address (usually 100) must also appear.

See

http://www.geocities.com/thestarman3/asm/debug/debug.htm

or

http://www.ping.be/~ping0751/debug.htm

for more information on DEBUG.

How to unassemble with DEBUG

DEBUG is a very odd program! It can be used to look into RAM directly or into any file, it can assemble or unassemble files, it can run programs and be used to change as little as 1 bit.

It is started at the DOS prompt by

C:\>DEBUG yourfile.txt

and all you get is a '-' character!

This '-' character is the input prompt. Possible commands are shown below.

```
assemble        A [address]
compare         C range address
dump            D [range]
enter           E address [list]
fill            F range list
go              G [=address] [addresses]
hex             H value1 value2
input           I port
load            L [address] [drive] [firstsector] [number]
move            M range address
name            N [pathname] [arglist]
output          O port byte
proceed         P [=address] [number]
quit            Q
register        R [register]
search          S range list
trace           T [=address] [value]
unassemble      U [range]
write           W [address] [drive] [firstsector] [number]
allocate expanded memory         XA [#pages]
deallocate expanded memory       XD [handle]
map expanded memory pages        XM [Lpage] [Ppage] [handle]
display expanded memory status   XS
```

For the use that DEBUG is put to here, to unassemble a program, the only commands needed are U for Unassemble and Q for Quit.

To unassemble a file called TRP1.COM simply issue the command

C:\> DEBUG TRP1.COM

at the command prompt then use the command U (then enter). You can unassemble from any address and any size, the command U 100 1000 unassembles from address 100 and 1000 bytes long, both numbers in hex.

If you need the unassembly in a file, you should first prepare a command file that contains just the data below. Assume this command file is called UNASSM.DAT, it is made using the commands at the DOS prompt

C:\> COPY CON UNASSM.DAT
U
Q
CTRL Z

where CTRL Z means press the CTRL and Z keys together.

You can now produce the unassembled .com file in another file by using I/O redirection as in the command:

C:\> DEBUG TRP1.COM < UNASSM.DAT > TRP1.DAT

The < character means take the input from your command file called UNASSM.DAT and the > character means place or redirect the output into whatever file name you provide, in this case TRP1.DAT.

If you need to unassemble more than 100 hex bytes, the file UNASSM.DAT will have to contain

U 100 200
Q
CTRL Z

where the 200 refers to 200 hex bytes in length. If you need to quote a length in the U command, the address 100 must also appear.

See

http://www.geocities.com/thestarman3/asm/debug/debug.htm

or

http://www.ping.be/~ping0751/debug.htm

for more information on DEBUG.

```
;PRINTSTRING prints a string pointed to with SI register
;OUTINT writes a 16 bit number in AX to the screen as integer
;PRINTNUM   writes a string as PRINTSTRING but with no spaces
;SHOWBITS   writes a 16 bit number in AX to the screen as binary
;leadingzero       writes a single 0 to the screen
;crlf writes an ASCII 13 Carriage Return then ASCII 10 Line Feed


rem       dw 0               ;16 bit variable for remainder
quo       dw 0               ;16 bit variable for quotient
space             equ " "    ;defines "space" as ASCII 32
numbuff:  db 5   dup(space)  ;5 duplicates of "space"
          db 0               ;to give zero byte string
                             ;terminator
row       db 0               ;for MOVECURSOR
col       db 0               ;for MOVECURSOR



;.......... Subroutines ..............................



printstring:        ;assume pointer to
                    ;string is in SI register
     mov ah,02
L1:  mov dl,[si] ;get first character
     cmp dl,0    ;see if end of string
     jz finprint ;finish if end of string
     int 021     ;write char
     inc si      ;point to next char
     jmp L1      ;back for more chars
finprint:           ret ;end of subroutine

;...... end of printstr .....................
```

```
;  subroutine  to  show  a  16  bit  number  in  AX  as  binary


showbits:
          mov  si,ax      ;copy  of  equipment  config  number
          mov  bl,16      ;counter  for  16  bits
L1:       mov  cx,si      ;cx  is  working  register
          and  cx,08000   ;mask  off  all  but  top  bit
          cmp  cx,0       ;if  zero
          jz  nought      ;then  write  a  '0'


one:      mov  dl,'1'
          mov  ah,02      ;DOS  write  to  screen  function
          int  021
          jmp  next


nought:   mov  dl,'0'
          mov  ah,02      ;DOS  write  to  screen  function
          int  021


next:     shl  si,1       ;move  all  bits  left  one  space
          dec  bl         ;
          jnz  L1         ;see  if  all  16  bits  done
          ret             ;end  of  showbits  subroutine


;.............end  if  showbits...................


;  OUTINT  subroutine
;this  prints  out  a  number  in  decimal  held  in  AX  register  by     ;storing  it
;as  ascii  into  a  buffer  and  then  pointing  to  this  buffer  with  ;bx.
;Version  2,  max  number  now  65535  (was  255)
;Needs  numbuff  and  space  equ  at  top  of  file.  A86  seems  to  fall  ;over
;if  these  are  included  within  the  code.


outint:
          mov  cx,5            ;for  each  place  in  the  buffer
          mov  si,numbuff      ;
          mov  dl,space        ;
L21:      mov[si],dl           ;fill  it  with  spaces
          inc  si
          loop  L21            ;the  LOOP  instruction
                               ;uses  CX  and  does  the
                               ;increment  for  you  first
          mov  si,numbuff +4   ;ready  to  start  filling  the  array



L22:
          mov  dx,0       ;clear  DX  since  DIV  works  on  DX:AX
          mov  cx,10      ;divisor  of  10
          div  cx         ;divide  by  10  and  store  remainder
```

```
                mov  quo,ax       ;store quotient for next time
                or   dl,48        ;change remainderto ascii
                mov  [si],dl      ;and store in buffer
                dec  si           ;point to next location in buffer
                mov  ax,quo       ;ready for next div
                cmp  ax,0         ;see if any numbers left
                jne  L22          ;if there are, process them
                                  ;if not, write result to screen
                mov  si,numbuff   ;point to the buffer
                call printnum     ;output the result

                ret               ;return to whence you came

printnum:                         ;assume pointer to string
                                  ;is in SI register
                mov  ah,02
L51:            mov  dl,[si]      ;get first character
                cmp  dl,0         ;see if end of string
                jz   finprintnum  ;finish if end of string
                cmp  dl,32        ;see if space char
                jz   L52          ;do not print spaces
                int  021          ;write char
L52:            inc  si           ;point to next char
                jmp  L51          ;back for more chars

finprintnum:                      ret

;............end if outint.................

;............crlf subroutine.................

crlf:    pusha              ;save gen purpose registers
         mov  ah,2          ;DOS write char function
         mov  dl,13         ;Carriage Return (CR)
         int  021          ;print it
         mov  dl,10         ;Line Feed (LF)
         int  021          ;print it
         popa              ;restore registers
         ret               ;return from subroutine

leadingzero:
         PUSH AX            ;keep AX to prevent overwriting
         MOV  AH, 2         ;DOS int 20 function 2
         MOV  DL, "0"       ;zero character to be output
         INT  021          ;call DOS
         POP  AX            ;restore value to AX
         RET               ;return to address after calling address
```

Answers

Unit 2

Worksheet 1

Signed	Unsigned
183	183
89	89
23912	23912
−349	3117
9481	9481
−7399	40167

Worksheet 2

Signed	Unsigned
28550	28550
29374	29374
15316	15316
29187	29187
26692	26692
4197	4197

Worksheet 3

Signed	Unsigned
5632	5632
14286	14286
10624	10624
8728	8728
13022	13022
26001	26001

Worksheet 4

Signed	Unsigned
28748	28748
11082	11082
26044	26044
18740	18740
26218	26218
8733	8733

Worksheet 5

Signed	Unsigned
22807	22807
25774	25774
10294	10294
7789	7789
3632	3632
28705	28705

Worksheet 6

Signed	Unsigned
−29241	62009
−25401	58169
15219	15316
15408	29187
−714	33482
−12406	45174

Worksheet 7

Signed	Unsigned
−13334	46102
6693	6693
1932	1932
−14691	47459
19677	19677
−17121	49889

Worksheet 8

Signed	Unsigned
29068	29068
19032	19032
21132	21132
−29789	62557
23544	23544
−14835	47603

Worksheet 9

Signed	Unsigned
10347	10347
25855	25855
4907	4907
−7140	39908
−17751	50519
42	42

Worksheet 10

Signed	Unsigned
−29568	62336
24323	24323
5883	5883
1823	1823
28178	28178
684	684

Worksheet 11

	Hex values, little endian	Hex values, big endian	Decimal value
Image file size	36 1B B7 00	B7 1B 36	12 000 054
Image width	D0 07 00 00	7D0	2000
Image height	D0 07 00 00	7D0	2000
Number of bits per pixel	18 00	18	24
Max. number of colours possible	n/a	n/a	16.7 million

Worksheet 12

	Hex values, little endian	Hex values, big endian	Decimal value
Image file size	8E 37 38 00	38378E	3 684 238
Image width	22 09 00 00	922	2338
Image height	26 06 00 00	626	1574
Number of bits per pixel	08 00	8	8
Max. number of colours possible	n/a	n/a	256

Worksheet 13

	Hex values, little endian	Hex values, big endian	Decimal value
Image file size	36 96 10 00	109636	1 087 030
Image width	C4 04 00 00	4C4	1220
Image height	F0 06 00 00	6F0	1776
Number of bits per pixel	04 00	4	4
Max. number of colours possible	n/a	n/a	16

Worksheet 14

	Hex values, little endian	Hex values, big endian	Decimal value
Image file size	62 D6 01 00	1D662	120 418
Image width	52 01 00 00	152	338
Image height	5F 01 00 00	15F	351
Number of bits per pixel	08 00	8	8
Max. number of colours possible	n/a	n/a	256

Worksheet 15

	Hex values, little endian	Hex values, big endian	Decimal value
Image file size	36 50 1D 00	1D5036	1 921 078
Image width	40 06 00 00	640	1600
Image height	B0 04 00 00	4B0	1200
Number of bits per pixel	08 00	8	8
Max. number of colours possible	n/a	n/a	256

Worksheet 16

	Hex values, little endian	Hex values, big endian	Decimal value
Image file size	C2 86 20 00	2086C2	2 131 650
Image width	CB 03 00 00	3CB	971
Image height	DB 02 00 00	2DB	731
Number of bits per pixel	18 00	18	24
Max. number of colours possible	n/a	n/a	16.7 million

Worksheet 17

Decimal	Hex	Octal	Binary
117 212	1C9DC	344734	0001110010011101
1431	597	2627	010110010111
1044	414	2024	010000010100
4	4	4	0100
91 509	16575	262565	0001011001010111
1248	4E0	2340	010011100000
69	45	105	01000101
100 875	18A0B	305013	0001100010100000
340	154	524	000101010100
66 182	10286	201206	0001000000101000
8712	2208	21010	0010001000001000
10 815	2A3F	25077	0010101000111111

Worksheet 18

Decimal	Hex	Octal	Binary
70 149	11205	211005	0001000100100000
4252	109C	10234	0001000010011100
92 663	169F7	264767	0001011010011111
47	2F	57	00101111
10 229	27F5	23765	0010011111110101
232	E8	350	11101000
589	24D	1115	001001001101
26 805	68B5	64265	0110100010110101
221	DD	335	11011101
3160	C58	6130	110001011000
8813	226D	21155	0010001001101101
445	1BD	675	000110111101

Worksheet 19

Decimal	Hex	Octal	Binary
10 272	2820	24040	0010100000100000
4599	11F7	10767	0001000111110111
10 308	2844	24104	0010100001000100
8246	2036	20066	0010000000110110
377	179	571	000101111001
2947	B83	5603	101110000011
6443	192B	14453	0001100100101011
11 389	2C7D	26175	0010110001111101
9148	23BC	21674	0010001110111100
10 332	285C	24134	0010100001011100
127 139	1F0A3	370243	0001111100001010
3479	D97	6627	110110010111

Worksheet 20

Decimal	Hex	Octal	Binary
18 318	478E	43616	0100011110001110
7948	1F0C	17414	0001111100001100
8326	2086	20206	0010000010000110
6443	192B	14453	0001100100101011
1025	401	2001	010000000001
93 208	16C18	266030	0001011011000001
98 757	181C5	300705	0001100000011100
4734	127E	11176	0001001001111110
55 845	DA25	155045	1101101000100101
11 453	2CBD	26275	0010110010111101
120 084	1D514	352424	0001110101010001
210	D2	322	11010010

Worksheet 21

Decimal	Hex	Octal	Binary
350	15E	536	000101011110
663	297	1227	001010010111
510	1FE	776	000111111110
1706	6AA	3252	011010101010
8742	2226	21046	0010001000100110
72 928	11CE0	216340	0001000111001110
8702	21FE	20776	0010000111111110
8458	210A	20412	0010000100001010
12 469	30B5	30265	0011000010110101
4272	10B0	10260	0001000010110000
77	4D	115	01001101
5474	1562	12542	0001010101100010

Worksheet 22

Decimal	Hex	Octal	Binary
12 423	3087	30207	0011000010000111
8683	21EB	20753	0010000111101011
74 677	123B5	221665	0001001000111011
112 061	1B5BD	332675	00011011010110111101
359	167	547	000101100111
233	E9	351	11101001
2973	B9D	5635	101110011101
123 366	1E1E6	360746	0001111000011110
29	1D	35	00011101
10 778	2A1A	25032	0010101000011010
786	312	1422	001100010010
10 167	27B7	23667	0010011110110111

Worksheet 23

Answer			A AND B	B XOR C	D OR E
A	B	C	D	E	R
0	0	0	0	0	0
0	0	1	0	1	1
0	1	0	0	1	1
0	1	1	0	0	0
1	0	0	0	0	0
1	0	1	0	1	1
1	1	0	1	1	1
1	1	1	1	0	1

Worksheet 24

Answer			A OR B	B NAND C	D OR E
A	B	C	D	E	R
0	0	0	0	1	0
0	0	1	0	1	0
0	1	0	1	1	1
0	1	1	1	0	0
1	0	0	1	1	1
1	0	1	1	1	1
1	1	0	1	1	1
1	1	1	1	0	0

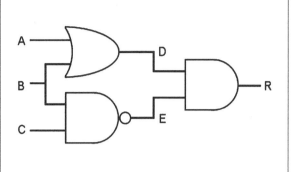

Worksheet 25

Answer			A NAND B	B NOR C	D XOR E
A	B	C	D	E	R
0	0	0	1	1	0
0	0	1	1	0	1
0	1	0	1	0	1
0	1	1	1	0	1
1	0	0	1	1	0
1	0	1	1	0	1
1	1	0	0	0	0
1	1	1	0	0	0

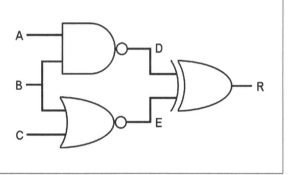

Worksheet 26

Answer			A AND B	B NOR C	D OR E
A	B	C	D	E	R
0	0	0	0	1	1
0	0	1	0	0	0
0	1	0	0	0	0
0	1	1	0	0	0
1	0	0	0	1	1
1	0	1	0	0	0
1	1	0	1	0	1
1	1	1	1	0	1

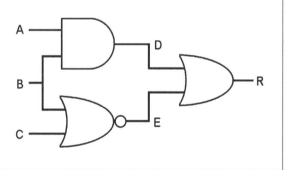

Worksheet 27

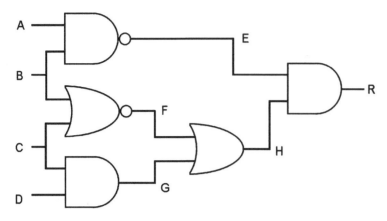

Answer				A NAND B	B NOR C	C AND D	F OR G	E AND H
A	B	C	D	E	F	G	H	R
0	0	0	0	1	1	0	1	1
0	0	0	1	1	1	0	1	1
0	0	1	0	1	0	0	0	0
0	0	1	1	1	0	1	1	1
0	1	0	0	1	0	0	0	0
0	1	0	1	1	0	0	0	0
0	1	1	0	1	0	0	0	0
0	1	1	1	1	0	1	1	1
1	0	0	0	1	1	0	1	1
1	0	0	1	1	1	0	1	1
1	0	1	0	1	0	0	0	0
1	0	1	1	1	0	1	1	1
1	1	0	0	0	0	0	0	0
1	1	0	1	0	0	0	0	0
1	1	1	0	0	0	0	0	0
1	1	1	1	0	0	1	1	0

Worksheet 28

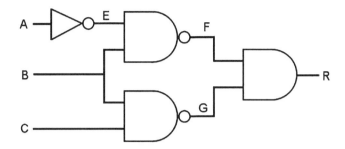

Answer			NOT A	D NAND B	E NAND B	B NAND C	F AND G
A	B	C	D	E	F	G	R
0	0	0	1	1	1	1	1
0	0	1	1	1	1	1	1
0	1	0	1	0	1	1	1
0	1	1	1	0	1	0	0
1	0	0	0	1	1	1	1
1	0	1	0	1	1	1	1
1	1	0	0	1	0	1	0
1	1	1	0	1	0	0	0

Worksheet 29

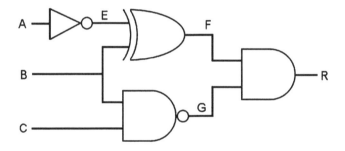

Answer			NOT A	D NAND B	**E XOR B**	B NAND C	F AND G
A	B	C	D	E	**F**	G	R
0	0	0	1	1	**1**	1	1
0	0	1	1	1	**1**	1	1
0	1	0	1	0	**1**	1	1
0	1	1	1	0	**1**	0	0
1	0	0	0	1	**1**	1	1
1	0	1	0	1	**1**	1	1
1	1	0	0	1	**0**	1	0
1	1	1	0	1	**0**	0	0

Worksheet 30

Possible functions to use in the spreadsheet

A	B	C	D
1	0.1	=	=CONCATENATE(B65,B66)
2			
3	=B1	=B3*2	=INT(C3)
4	=C3-D3	=B4*2	=INT(C4)
5	=C4-D4	=B5*2	=INT(C5)
6	=C5-D5	=B6*2	=INT(C6)
7	=C6-D6	=B7*2	=INT(C7)
8	=C7-D7	=B8*2	=INT(C8)
	Rows deleted to save space		
57	=C56-D56	=B57*2	=INT(C57)
58	=C57-D57	=B58*2	=INT(C58)
59	=C58-D58	=B59*2	=INT(C59)
60	=C59-D59	=B60*2	=INT(C60)
61	=C60-D60	=B61*2	=INT(C61)
62	=C61-D61	=B62*2	=INT(C62)
63			
64			
65	=CONCATENATE(".",D3,D4,D5,D6,D7,D8,D9,D10,D11,D12,D13,D14,D15,D16, D17,D18,D19,D20,D21,D22,D23,D24,D25,D26,D27,D28,D29,D30,D31,D32)		
66	=CONCATENATE(D33,D34,D35,D36,D37,D38,D39,D40,D41,D42,D43,D44,D45, D46,D47,D48,D49,D50,D51,D52,D53,D54,D55,D56,D57,D58,D59,D60,D61)		

Concatenate means 'join together'. Microsoft Excel can only join up to 30 items with the concatenate function so this spreadsheet uses two concatenate functions then joins the results.

Decimal	60 place binary	48-bit accurate
0.1	.000110011001100110011001100110011001100110011001100110011010000	No
0.25678	.010000011011110001010101100001100100010001010010010000000000	No
0.0625	.000100000000000000000000000000000000000000000000000000000000	Yes
0.251220703	.010000000101000000000000000000000000000000000000000000000000	Yes
0.141592654	.001001000011111101101010100010001000010110100011000000000000	Yes

Worksheet 31

Worksheet 31 possible answer version 1. This uses a modified version of the printstring routine from LIB.ASM.

```
        mov  si,buffer
                ;POINT to string in buffer with SI
                ;students should understand that the string is NOT
                ;put in SI, it is only a pointer
        call printstring
                ;output the string
        int 020     ;terminate program


;printstring routine copied from LIB.ASM then modified

printstring:
                ;assume pointer to string is in SI register
        mov ah,02    ;DOS Output character routine 2
L1:     mov dl,[si] ;get first character
        cmp dl,0     ;see if end of string
        jz finprint ;finish if end of string
        or dl,020    ;Modification to PRINTSTRING to convert to lower case
                     ;020 is 32 decimal and is a MASK
        int 021      ;write char
        inc si       ;point to next char
        jmp L1       ;back for more chars
finprint:   ret     ;terminate program


buffer:    db "This is a ASCIIZ STRING in MIXED CAPS and lowercase",0
```

Worksheet 31 possible answer version 2. This converts the string to lowercase in its buffer then uses the standard printstring routine from LIB.ASM.

```
;Program to output a string in a buffer all in lowercase

        mov si,buffer      ;POINT to start of string buffer
L1:     mov dl,[si]        ;get byte from buffer
        cmp dl,0           ;check for end of ASCIIZ string
        je output          ;if end, go to output section
        or dl,020          ;convert to lowercase with 020 (32 decimal) MASK
        mov [si],dl        ;put back in the buffer in the same place
        inc si             ;point to next location in buffer
        jmp L1             ;and go back for some more

output:
        mov si,buffer      ;POINT to start of buffer again
        call printstring   ;output the buffer as string
        int 020            ;terminate program


buffer:    db "This is a ASCIIZ STRING in MIXED CAPS and lowercase",0
```

Worksheet 32

Program to read two numeric keypresses and display their sum. Keypresses have to be stored because they will be overwritten by the DOS routines.

```
mov ah,1     ;DOS read keyboard function number 1
int 021      ;call DOS
mov ch,al    ;store key to prevent being overwritten
mov ah,2     ;DOS output character function number 2
mov dl,"+"   ;character to be output
int 021      ;call DOS
mov ah,1     ;DOS read keyboard function number 1
int 021      ;call DOS
mov cl,al    ;store key to prevent being overwritten
mov ah,2     ;DOS output character function number 2
mov dl,"="   ;character to be output
int 021      ;call DOS
and ch,0CF   ;convert first digit ASCII to decimal
and cl,0CF   ;convert second digit ASCII to decimal
xor ax,ax    ;clear ax, i.e. ah and al ready for addition
add al,ch    ;add first digit, could have used mov al,ch
add al,cl    ;add second digit
call outint  ;output multi digit answer
int 020      ;terminate program
```

Worksheet 33

```
;TRP worksheet 203 answer
;program to output only characters in the range a-z
;all spaces and other characters to be removed

        mov si,buffer       ;POINT to string in buffer with SI
                            ;students should understand that the string is NOT
                            ;put in SI, it is only a pointer
        call printstring
        int 020             ;terminate program


;printstring routine copied from LIB.ASM then modified

printstring:                ;assume pointer to string is in SI register
        mov ah,02           ;DOS Output character routine 2
L1:     mov dl,[si]         ;get first character
        cmp dl,0            ;see if end of string
        jz finprint         ;finish if end of string

        cmp dl,"a"          ;section to "jump over" the int 021 to
                            ;avoid printing characters
        jl L2
        cmp dl,"z"
        ja L2

        int 021             ;write char
L2:     inc si              ;point to next char
        jmp L1              ;back for more chars

finprint:
        ret                 ;terminate program

buffer:    db "This is a ASCIIZ STRING in MIXED CAPS and lowercase",0
```

Worksheet 34

First version with a bug:

This program 'looks' like it should work. It reads the keyboard OK and the output works but whatever key is pressed, it outputs the same value. This is because the value gets overwritten by the DOS function 2. The solution is to store the keypress then restore it just when it is required. This is shown in the second version.

```
;program to read a key and display its ASCII value

        mov ah,1    ;DOS read keyboard function number 1
        int 021     ;call DOS
        mov ah,2    ;DOS output character function 2
        mov dl,"="  ;character to output
        int 021     ;output
        xor ah,ah   ;clear ah register to 0 (could use mov ah,0)
        call outint ;output multidigit number
        int 020     ;terminate program
```

Second version:

```
;program to read a key and display its ASCII value

        mov ah,1    ;DOS read keyboard function number 1
        int 021     ;call DOS
        mov cl,al   ;store keyvalue to avoid overwriting
        mov ah,2    ;DOS output character function 2
        mov dl,"="  ;character to output
        int 021     ;output
        xor ah,ah   ;clear ah register to 0 (could use mov ah,0)
        mov al,cl   ;restore original keypress
        call outint ;output multidigit number
        int 020     ;terminate program
```

Worksheet 35

This program uses a binary mask to mask off or set to zero all the bits in the AL register except the one that hold the status of the num lock, i.e. bit 5. As $2^5 = 32$, the required value of the mask is 32. The AND operation forces the other bits to zero. The program could be easily extended to cover CAPS lock, Scroll lock etc.

```
;program to show if the numlock is on or off

        mov  ah,012  ;BIOS interupt 16 function 12, display keyboard status
        int  016     ;call BIOS
        and  al,32   ;clear all return values except bit 5 by masking
        jz L2        ;if masking made al=0, the numlock is off
        mov  si, numlockon
;display "on" string

        call printstring
        int  020     ;terminate program


L2:   mov  si,numlockoff
;display "off" string
        call printstring
        int  020     ;terminate program



;Define strings

numlockon:
        db "The numlock is on",0
numlockoff:
        db "The numlock is off",0
```

Unit 4

Worksheet 1

```
program worksht one (input;output);
      {space in program name, ; instead of , in input,output }

var;x,y,z:integer;
      { ; instead of : after var }

      earnings,tax,x:real;
      { x defined as variable but name x has been used already}

      st,for,name:string
      {no ; at the end of the statement}
      {reserved word "for" used as var name }

begin
      write("What is your name ");
      { wrong quote marks, should be ' ' }

      readln(yourname);
      { variable not defined }

      writeln('Hello ',name, ', how much to do earn each year? );
      { missing quote after year? }

      readline(earnings);
      { should be readln not readline }

      tax=earning*0.22;
      { variable earning not defined, should be earnings }

      writeline(You will pay', tax:0:3,' pounds tax');
      { writeline should be writeln }

end; { end of program should be end. not end; }
```

Worksheet 2

```
program  2ndversion_of_table;
        { cannot start an identifier with a number (2) }
var  counter:integer;
      i:real;


function  getnumber(lower,upper:integer):integer;
var  inputstr:string;
      errorcode:integer;
      x:integer;

begin
      while
      {while should be repeat }
            write('Type a value between ',lower,' and
                        ',upper,' inclusive');
            readln(inputstr);
            val(inputstr,x,errorcode);
      until ((errorcode=0) and (x>=lower) and (x<=upper));
    { function value must be assigned to function name with getnumber:=x; )
end;


begin   {start of main program block }

      counter:=getnumber(2,12);
      for i:=1 to 12 do;
            {i is of var type real so cannot be used as a counter here }
            { the ; after do will stop the loop executing }

                  writeln(i,' times ',counter,' = ',i*counter);
end.
```

Worksheet 3

```
program area;

var  a,b:real;
     x,y:float;
          { float is not a Pascal variable type, it belongs in c }


procedure area(a,b:real);
          { identifier "area" has been used already as the program name,
          { identifiers must be unique }
begin
     var area,a,b:real;
          { var definition block must be outside execution block }
          { identifier "area" has been used already,
              identifiers must be unique }
     area=a*b;
          { assignment operator should be := }
end.
          { end of procedure definition should be end; }


begin
     print("Program to print area of rectangle");
          { print is a BASIC reserved word, should be write }
          { wrong quotes, should be ' ' }
     print("What is the length of side A? ");
          { print is a BASIC reserved word, should be write }
          { wrong quotes, should be ' ' }
     readln(a);
     print('What is the length of side B? );
          { print is a BASIC reserved word, should be write }
          { missing quote after B? }
     readln(b);
          { should be readln not read1n}
     area(a,b);
     print(The area is ',area:0:2);
          { print is a BASIC reserved word, should be write }
          { no way for a value to be returned, variable out of scope }
          { missing quote before The }
end;
          { end of program definition should be end. }
```

Worksheet 4

```
program conversion(output,input);
      { In standard Pascal, the program header ends in (input, output) }
      { (input,output) is optional in both Free Pascal and Borland
            Turbo Pascal }

uses crt;

function miles2km(x:integer);
      { no function return type, should be real }
start
      { should be begin not start }
      milestokm:=x/1.609344;
      { should be x*1.609344 not x/1.609344 }
      { user defined function name is miles2km not milestokm }
end;

begin
      write(Program to convert miles to kilometres');
      { missing quote at start of string }
      write('What is the distance in miles? ')
      { missing ; character at end of statement }
      readln(miles);
      { var not defined }
      writeln(miles,' equals ',miles2km(miles),' kilometres);
      { missing quote after kilometres }
      clrscr;
      { cause a blank screen so output will not be visible }
end.
```

Worksheet 5

The statement for `i:=1 to length(name) do;` has a ';' indicating end of statement but the real end of statement is the 'end;' directly below the 'begin' of the 'for' loop. The result is that no looping will occur. Pascal will count from 1 to the length of the string then execute what should have been looped but only once. The program with the fault in place will output a single character of the string only (in capitals) as the variable 'i' will not have been assigned an initial value.

```
for I:=1 to length(name) do;   { ; should not be here }
    begin
            ch:=copy(name,I,1); { get one char at a time }
            if convert_to_caps=true
                then
                begin
                        { add upper case char to output string }
                        outputstr:=concat(outputstr,upcase(ch));
                        { set boolean ready for next time}
                        convert_to_caps:=false;
                end
                else
                begin
                        { add lower case char to output string }
                        outputstr:=concat(outputstr,ch);
                end;

            { see if next char must be capital }
            if ch=' ' then convert_to_caps:=true;
    end; { end of "for" loop }
```

Worksheets 6 to 9

Using stepwise refinement, break down the problem into these parts:

- Get data, i.e. get room dimensions and paint coverage.
- Do calculation.
- Output the result.

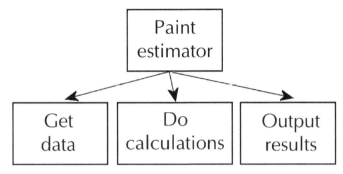

Figure 4.1 *Stepwise refinement*

Example test data without data validation tests:

Width of wall A	Width of wall B	Height	Paint coverage	Ceiling area	Wall area	Total area	Litres to paint walls only	Litres to paint all	Rounded litres to paint walls only	Rounded litres to paint all
3	2	2.5	15	6	25	31	1.67	2.07	2	3
5	3	3	15	15	48	63	3.20	4.20	4	5
1	1.5	2	15	1.5	10	11.5	0.67	0.77	1	1
1.5	4	2.7	17	6	30	35.7	1.75	2.10	2	3
2.9	2	2.5	15	5.8	25	30.3	1.63	2.02	2	2

Test sheet:

Inputs					Expected output Rounded litres	Actual ouput Rounded litres	OK Y or N
Width of wall A	Width of wall B	Height	Paint coverage	Paint ceiling?			
3	2	2.5	15	Y	3		
5	3	3	15	Y	5		
1	1.5	2	15	Y	1		
1.5	4	2.7	17	Y	3		
2.9	2	2.5	15	Y	2		
3	2	2.5	15	N	2		
5	3	3	15	N	4		
1	1.5	2	15	N	1		
1.5	4	2.7	17	N	2		
2.9	2	2.5	15	N	2		

First attempt at program (with a bug as will be shown below).

```
program trp101;

{Program instructions included here }
{The calculation is
((area of wall A*2) + (area of wall B*2)) / paint coverage.
If the ceiling is to be painted, use the wall dimensions to
calculate ceiling size then round up to the nearest litre.
Only round up total wall+ceiling paint required at the end
of the program to avoid 2 rounding errors.}

var  wallA,wallB,height,coverage:real;
         wallarea,totalarea,paintneeded:real;
      paintceiling:boolean;
      answer:char;


begin
      {start of getdata section}
      writeln('Paint coverage program version 1.0');
      writeln('Please input values in metres');
      write('What is the length of the first wall? ');
      readln(wallA);
      write('What is the length of the second wall? ');
      readln(wallB);
      write('What is the height of the room? ');
      readln(height);
      write('Do you wish to allow for painting the ceiling,
            type Y or N? ');
      readln(answer);
      write('What is the coverage of the paint
            in square metres per litre? ');
      readln(coverage);

      {start of calculation section }
      if (answer='Y') or (answer='y') then
            {allow any other key to be equal to "no"}

            paintceiling:=true
            else
            paintceiling:=false;

      wallarea:=(wallA*height*2)+(wallB*height*2);

      if paintceiling then
            totalarea:=wallarea+(wallA*wallB)
            else
            totalarea:=wallarea;
```

```
    paintneeded:=totalarea/coverage;

    {check for 2% tolerance in paint quantity}
        if paintneeded/int(paintneeded) >=1.02 then
        paintneeded:=int(paintneeded)+1
        else
        paintneeded:=int(paintneeded);


    {start of output section}
    write('Paint required is ',paintneeded:0:0,' litres');
    if paintceiling then
        write(' to cover walls and ceiling')
        else
        write(' to cover the walls only');
end.
```

Output using data from line 1 of the test data.

```
Paint coverage program version 1.0
Please input values in metres
What is the length of the first wall? 3
What is the length of the second wall? 2
What is the height of the room? 2.5
Do you wish to allow for painting the ceiling, type Y or N? y
What is the coverage of the paint in square metres per litre? 15
Paint required is 3 litres to cover walls and ceiling
```

From this data, the program seems to work but if all the lines in the test data are used, a problem arises as is shown in the test data table.

Inputs					Expected output Rounded litres	Actual ouput Rounded litres	OK Y or N
Width of wall A	Width of wall B	Height	Paint coverage	Paint ceiling?			
3	2	2.5	15	Y	3	3	Y
5	3	3	15	Y	5	5	Y
1	1.5	2	15	Y	1	Div by 0	N
1.5	4	2.7	17	Y	3	3	Y
2.9	2	2.5	15	Y	2	2	Y
3	2	2.5	15	N	2	2	Y
5	3	3	15	N	4	4	Y
1	1.5	2	15	N	1	Div by 0	N
1.5	4	2.7	17	N	2	2	Y
2.9	2	2.5	15	N	2	2	Y

Using data from the 3rd and 8th lines of the test data causes a program crash with 'Divide by zero'.

There are only two division operations in the program so the bug must be from one of these lines:

```
paintneeded:=totalarea/coverage;
or
if paintneeded/int(paintneeded)  >=1.02 then
```

The first one cannot generate a div by 0 error if the value of the variable `coverage` is not zero; the bug must be in the second division. This line is used in the rounding decision, i.e. if the amount is within 2% of an integer number of litres. In the case where the test data gave a div by 0 error, the value of `int(paintneeded)` is 0 as the amount of paint required is small. The solution is to test the value of `int(paintneeded)` before the division like this:

```
if int(paintneeded)>0 then
     if paintneeded/int(paintneeded)  >=1.02 then
          paintneeded:=int(paintneeded)+1
          else
          paintneeded:=int(paintneeded);

if int(paintneeded)=0 then paintneeded:=1;
```

so the complete program is now:

```
program trp101a;

{Program instructions included here }
{The calculation is
((area of wall A*2) + (area of wall B*2)) / paint coverage.
If the ceiling is to be painted, use the wall dimensions to
calculate ceiling size then round up to the nearest litre.
Only round up total wall+ceiling paint required at the end
of the program to avoid 2 rounding errors.}


var wallA,wallB,height,coverage:real;
      wallarea,totalarea,paintneeded:real;
    paintceiling:boolean;
    answer:char;

begin
     {start of getdata section}
     writeln('Paint coverage program version 1.0');
     writeln('Please input values in metres');
     write('What is the length of the first wall? ');
     readln(wallA);
     write('What is the length of the second wall? ');
     readln(wallB);
     write('What is the height of the room? ');
     readln(height);
```

```
write('Do you wish to allow for painting the ceiling,
      type Y or N? ');
readln(answer);
write('What is the coverage of the paint
      in square metres per litre? ');
readln(coverage);

{start of calculation section }
if (answer='Y') or (answer='y') then
      {allow any other key to be equal to "no"}
      paintceiling:=true
      else
      paintceiling:=false;

wallarea:=(wallA*height*2)+(wallB*height*2);

if paintceiling then
      totalarea:=wallarea+(wallA*wallB)
      else
      totalarea:=wallarea;

paintneeded:=totalarea/coverage;

{check for 2% tolerance in paint quantity}

      if int(paintneeded)>0 then   {line added as result of testing}
          if paintneeded/int(paintneeded) >=1.02 then
            paintneeded:=int(paintneeded)+1
      else
      paintneeded:=int(paintneeded);

  if int(paintneeded)=0 then   paintneeded:=1;
      {line added as result of testing}

{start of output section}
write('Paint required is ',paintneeded:0:0,' litres');
if paintceiling then
      write(' to cover walls and ceiling')
      else
      write(' to cover the walls only');
end.
```

Worksheet 7

Test sheet:

	Expected outputs				Actual outputs				OK Y/N
White	**10**	**5**	**1**		**10**	**5**	**1**		
				total £				total £	
1	0	0	1	4.99	0	0	1	4.99	Y
2	0	0	2	9.98	0	0	2	9.98	Y
3	0	0	3	14.97	0	0	3	14.97	Y
4	0	0	4	19.96	0	0	4	19.96	Y
5	0	1	0	13.49	0	1	0	13.49	Y
6	0	1	1	18.48	0	1	1	18.48	Y
7	0	1	2	23.47	0	1	2	23.47	Y
8	0	1	3	28.46	0	1	3	28.46	Y
9	0	1	4	33.45	0	1	4	33.45	Y
10	1	0	0	26.99	1	0	0	26.99	Y
15	1	1	0	40.48	1	1	0	40.48	Y
16	1	1	1	45.47	1	1	1	45.47	Y
21	2	0	1	58.97	2	0	1	58.97	Y
23	2	0	3	68.95	2	0	3	68.95	Y
50	3	0	0	80.97	3	0	0	80.97	Y
Colours	**10**	**5**	**1**		**10**	**5**	**1**		
1	0	0	1	6.49	0	0	1	6.49	Y
2	0	0	2	12.98	0	0	2	12.98	Y
3	0	0	3	19.47	0	0	3	19.47	Y
4	0	0	4	25.96	0	0	4	25.96	Y
5	0	1	0	20.99	0	1	0	20.99	Y
6	0	1	1	27.48	0	1	1	27.48	Y
7	0	1	2	33.97	0	1	2	33.97	Y
8	0	1	3	40.46	0	1	3	40.46	Y
9	0	1	4	46.95	0	1	4	46.95	Y
10	1	0	0	39.99	1	0	0	39.99	Y
15	1	1	0	60.98	1	1	0	60.98	Y
16	1	1	1	67.47	1	1	1	67.47	Y
21	2	0	1	86.47	2	0	1	86.47	Y
23	2	0	3	99.45	2	0	3	99.45	Y
50	3	0	0	119.97	3	0	0	119.97	Y

```
program trp102;

{As program TRP101a but with addition of paint costs}

var wallA,wallB,height,coverage:real;
    wallarea,totalarea,paintneeded:real;
    paintceiling:boolean;
    answer:char;
    {new variables needed}
    intpaintneeded:integer;
    size10tins,size5tins,size1tins:integer;
```

```
      cost:real;
      iswhite:boolean;
      whatcolour:char;

begin
      {start of getdata section}
      writeln('Paint coverage program version 1.0');
      writeln('Please input values in metres');
      write('What is the length of the first wall? ');
      readln(wallA);
      write('What is the length of the second wall? ');
      readln(wallB);
      write('What is the height of the room? ');
      readln(height);
      write('Do you wish to allow for painting the ceiling,
           type Y or N? ');
      readln(answer);
      write('What is the coverage of the paint in square
           metres per litre? ');
      readln(coverage);
      write('What colour is required, W for white, C for colour ');
      readln(whatcolour);

      {start of calculation section }

      {allow any other key to be equal to "no"}
      if (answer='Y') or (answer='y') then paintceiling:=true
           else
           paintceiling:=false;

      wallarea:=(wallA*height*2)+(wallB*height*2);

      if paintceiling then
           totalarea:=wallarea+(wallA*wallB)
           else
           totalarea:=wallarea;

      paintneeded:=totalarea/coverage;

      {check for 2% tolerance in paint quantity}

      if int(paintneeded)>0 then
               if paintneeded/int(paintneeded) >=1.02 then
                       paintneeded:=int(paintneeded)+1
                       else
                       paintneeded:=int(paintneeded);
      if int(paintneeded)=0 then   paintneeded:=1;
           {line added as result of testing to cope with small amounts}

      {now calculate costs}
```

```
intpaintneeded:=trunc(paintneeded);
 {the trunc function will convert the real type paintneeded to
 an integer type intpaintneeded. In Pascal, you must be very
 careful with variable types. }


{use of integer arithmetic DIV and MOD to calculate number of tins}
{ DIV will give the number of tins and MOD gives the remainder
after using DIV ready to find the number of the next smaller size }

size10tins:= intpaintneeded DIV 10;
intpaintneeded:=intpaintneeded MOD 10;

size5tins:= intpaintneeded DIV 5;
intpaintneeded:=intpaintneeded MOD 5;

size1tins:= intpaintneeded;

        if (whatcolour='W') or (whatcolour='w') then
                  iswhite:=true
             else
                  iswhite:=false;

        if iswhite
             then
             cost:=(size10tins*26.99)+(size5tins*13.49)
             +(size1tins*4.99)
        else
             cost:=(size10tins*39.99)+(size5tins*20.99)
             +(size1tins*6.49);

        {start of output section}
        write('Paint required is ',paintneeded:0:0,' litres');
        if paintceiling then
             write(' to cover walls and ceiling')
             else
             write(' to cover the walls only');

        writeln;
        if size10tins>0 then {to avoid writing 0 tins which looks ugly}
             writeln('You will need ',size10tins,' 10 litre tins');

        if size5tins>0 then
             writeln('You will need ',size5tins,' 5 litre tins');

        if size1tins>0 then
             writeln('You will need ',size1tins,' 1 litre tins');

             writeln('The total cost will be ',cost:0:2,' pounds');

end.
```

Worksheet 8

What advantage does the use of procedures give over a single piece of program code regarding:

- The user
  None except that the program is more likely to be free of errors as a result of better program design and coding.
- The programmer
  For programs larger than a few lines, the use of procedures enables the programmer to think more clearly. If the overall programming task is large, each part may be written and tested in isolation. It is also easier to translate the progrem design into procedures as most design techniques assume that code is split into manageable parts.
- The programmer's employer
  For larger programs, the management can ensure the programming team have specialists that may be asked to complete certain parts of the project. For example, a specialist in screen layout could design the screen and a specialist in accounting would ensure the costings are accurate.

What disadvantage occurs when using only global variables?

Especially with reference to the answers above, when programs are written by teams of people, it is clearly impractical to ensure they each have their own set of global variable names to ensure one section does not clash with another. A local variable ensures that its value will not cause unforeseen problems in other sections of the whole program.

The following program is the same as trp102 except that the program is split into procedures. These have a direct relationship with the structure as described in the program design using stepwise refinement.

Only global variables have been used.

```
program trp103;

{As  program  TRP102  but  with  procedures}

var  wallA,wallB,height,coverage:real;
           wallarea,totalarea,paintneeded:real;
     paintceiling:boolean;
     answer:char;
     {new  variables  needed}
     intpaintneeded:integer;
     size10tins,size5tins,size1tins:integer;
     cost:real;
     iswhite:boolean;
     whatcolour:char;

procedure getdata;
begin
     {start  of  getdata  section}
     writeln('Paint  coverage  program  version  1.0');
     writeln('Please  input  values  in  metres');
```

```
      write('What is the length of the first wall? ');
      readln(wallA);
      write('What is the length of the second wall? ');
      readln(wallB);
      write('What is the height of the room? ');
      readln(height);
      write('Do you wish to allow for painting the ceiling,
            type Y or N? ');
      readln(answer);
      write('What is the coverage of the paint in
            square metres per litre? ');
      readln(coverage);
      write('What colour is required, W for white, C for colour ');
      readln(whatcolour);
end;
```

....

```
procedure calculate_paint_needed;

begin
      {allow any other key to be equal to "no"}
      if (answer='Y') or (answer='y') then paintceiling:=true
            else
            paintceiling:=false;

      wallarea:=(wallA*height*2)+(wallB*height*2);

      if paintceiling then
            totalarea:=wallarea+(wallA*wallB)
            else
            totalarea:=wallarea;

      paintneeded:=totalarea/coverage;

      {check for 2% tolerance in paint quantity}

      if int(paintneeded)>0 then
                  if paintneeded/int(paintneeded)  >=1.02 then
                        paintneeded:=int(paintneeded)+1
                        else
                        paintneeded:=int(paintneeded);

      if int(paintneeded)=0 then   paintneeded:=1;
            {line added as result of testing to cope with small amounts}

end;
```

....

```
procedure calc_cost;

    intpaintneeded:=trunc(paintneeded);
     {the trunc function will convert the real type paintneeded to
     an integer type intpaintneeded. In Pascal, you must be very
     careful with variable types. }


    {use of integer arithmetic DIV and MOD to calculate number of tins}
    { DIV will give the number of tins and MOD gives the remainder
    after using DIV ready to find the number of the next smaller size }

    size10tins:= intpaintneeded DIV 10;
    intpaintneeded:=intpaintneeded MOD 10;

    size5tins:= intpaintneeded DIV 5;
    intpaintneeded:=intpaintneeded MOD 5;


    size1tins:= intpaintneeded;

        if (whatcolour='W') or (whatcolour='w') then
                    iswhite:=true
             else
                    iswhite:=false;

        if iswhite
             then
             cost:=(size10tins*26.99)+(size5tins*13.49)
             +(size1tins*4.99)
        else
             cost:=(size10tins*39.99)+(size5tins*20.99)
             +(size1tins*6.49);

end;

..............................................................

procedure output_results;

    write('Paint required is ',paintneeded:0:0,' litres');
    if paintceiling then
        write(' to cover walls and ceiling')
        else
        write(' to cover the walls only');

    writeln;
    if size10tins>0 then   {to avoid writing 0 tins which looks ugly}
        writeln('You will need ',size10tins,' 10 litre tins');
```

```
        if size5tins>0 then
            writeln('You will need ',size5tins,' 5 litre tins');

        if size1tins>0 then
            writeln('You will need ',size1tins,' 1 litre tins');

            writeln('The total cost will be ',cost:0:2,' pounds');

end;

.............................................................

begin {start of main program }

        getdata;
        calculate_paint_needed;
        calc_cost;
        output_results;
end.
```

T - #0701 - 101024 - C0 - 297/210/11 - PB - 9780750656870 - Gloss Lamination